SYMBOLIC COMPUTATION

Artificial Intelligence

R.S. Michalski, J.G. Carbonell, T.M. Mitchell (Eds.): Machine Learning. An Artificial Intelligence Approach. XI, 572 pages, 1984

A. Bundy (Ed.): Catalogue of Artificial Intelligence Tools. Second, revised edition. IV, 168 pages, 1986

C. Blume, W. Jakob: Programming Languages for Industrial Robots. XIII, 376 pages, 145 figs., 1986

J.W. Lloyd: Foundations of Logic Programming. Second, extended edition. XII, 212 pages, 1987

L. Bolc (Ed.): Computational Models of Learning. IX, 208 pages, 34 figs., 1987

L. Bolc (Ed.): Natural Language Parsing Systems. XVIII, 367 pages, 151 figs., 1987

N. Cercone, G. McCalla (Eds.): The Knowledge Frontier. Essays in the Representation of Knowledge. XXXV, 512 pages, 93 figs., 1987

G. Rayna: REDUCE. Software for Algebraic Computation. IX, 329 pages, 1987

D.D. McDonald, L. Bolc (Eds.): Natural Language Generation Systems. XI, 389 pages, 84 figs., 1988

L. Bolc, M.J. Coombs (Eds.): Expert System Applications. IX, 471 pages, 84 figs., 1988

C.-H. Tzeng: A Theory of Heuristic Information in Game-Tree Search. X, 107 pages, 22 figs., 1988

L. Kanal, V. Kumar (Eds.): Search in Artificial Intelligence. X, 482 pages, 67 figs., 1988

H. Coelho, J. Cotta: Prolog by Example. 304 pages, 62 figs., 1988

H. Abramson, V. Dahl: Logic Grammars. XV, 258 pages, 40 figs., 1989

Harvey Abramson   Veronica Dahl

Logic Grammars

With 40 Illustrations

Springer Science+Business Media, LLC

Harvey Abramson
Department of Computer Science
University of Bristol
Queen's Building
Bristol BS8 1TR
England

Veronica Dahl
Department of Computer Science
Simon Fraser University
Burnaby, British Columbia
Canada V5A 1S6

Library of Congress Cataloging-in-Publication Data
Abramson, Harvey.
    Logic grammars.
    (Symbolic computation. Artificial intelligence)
    Includes bibliographies.
    1. Logic programming.   2. Artificial intelligence.
I. Dahl, Veronica, 1950–   . II. Title. III. Series.
QA76.6.A268 1989   006.3   88-35636

Printed on acid-free paper.

© 1989 by Springer Science+Business Media New York
Originally published by Springer-Verlag New York Inc. in 1989

Text prepared by authors in camera-ready form.

9 8 7 6 5 4 3 2 1

ISBN 978-1-4612-8188-7   ISBN 978-1-4612-3640-5 (eBook)
DOI 10.1007/978-1-4612-3640-5

To Lynn, and also to Cali who helped with the reading.

Harvey

To my mother, Selvita, and to Alice in Wonderland, for having provided role models with which it was possible to identify.

To Alexander and to Rob, for restoring the music in my heart.

Veronica

ACKNOWLEDGMENTS

I would like to ackowledge the influence of all the people who have stimulated my orientation toward logic, linguistics, and computational linguistics:

1. The phonetician Ivar Dahl, from whom I inherited or acquired a passion for the study of language.

2. The linguists Gabriel Bès, Alfredo Hurtado, Celia Jacubowicz, Beatriz Lavandera, Ana Maria Nethol, and Luis Prieto, whom in the precariousness of the Pampas developed my interest in formal linguistics—especially Alfredo Hurtado, with whom in exile I later started a collaboration that led to my research on Static Discontinuity for Government-Binding theory.

3. Alain Colmerauer and his group at Luminy, who introduced me to the marvels of logic programming and logic grammars and for more than two years provided me with a wonderfully friendly working atmosphere.

4. The veteran logic programming community, with whose lovely people from all countries I have always felt at home.

I would also like to thank Roland Sambuc, for our joint work on the first logic programmed expert system; Michael McCord, for our joint work on coordination and on discontinuous grammars; and all the people who, as visitors or members of my research group at Simon Fraser University, have contributed in one way or another to the work described in chapter 10: Michel Boyer, Charles Brown, Sharon Hamilton, Diane Massam, Pierre Massicotte, Brenda Orser, T. Pattabhiraman, Fred Popowich and Patrick Saint-Dizier.

Finally, I thank France, for having provided in 1975 the scholarship without which this book would never have been written, and Canada, my beautiful land of adoption. And, with special warmth, my many Latin American siblings, with whom I developed the resilience to keep aiming for the impossible.

Veronica Dahl

I would like to thank the following people and institutions for their help and or inspiration:

Ray Reiter, who while he was at the University of British Columbia, helped to de-isolate the place by arranging visits of some of the leading logic programmers and thus making it possible to get involved in the field at an early stage.

Alan Robinson, for his early encouragement of work which led to HASL and eventually to my work with grammars and language implementation. His combination of science, humanism, and generosity is exemplary.

David Turner for the example of SASL whose implementation in logic was so easy because the language was so elegant.

Professors Mike Rogers and John Shepherdson, and all the members of the Computer Science Department at the University of Bristol for the warmth of their reception and their interest while I spent a six month's sabbatical leave there from January to June of 1987.

Seif Haridi and the Logic Programming Group of the Swedish Institute of Computer Science and the chance to spend a few weeks there in a fine working environment.

Finally, the following three great things at UBC which helped to soothe the savaged academic:

1. The Wilson Recordings Library for providing music.

2. The Asian Studies Centre and Continuing Education for providing the opportunity of beginning to learn Japanese, and for providing a window to oriental languages and cultures.

3. And most of all ... (details on request).

Harvey Abramson

JOINT ACKNOWLEDGMENTS

We thank the students at Simon Fraser University and the University of British Columbia who helped us refine earlier versions of the material in this book. We are very grateful to Julie Johnson, G.M. Swinkels, and Bruce Weibe who spent so much time and energy in typesetting the book. Thanks to Barry J. Evans for a careful proof-reading of a near final version of the book.

We wish to thank the referees, whose pertinent comments were extemely valuable in improving the book. We also thank Springer-Verlag for its tolerance of our "fuzzy" interpretation of deadlines.

This book was completed with help to both of us from Canada's Natural Science and Engineering Research Council. V. Dahl's work described in section 3 of chapter 10 was partially supported by an S.U.R. research contract from IBM Canada.

Another such contract with H. Abramson, although not contributing material itself, did provide some support during the time of writing the book.

We wish to thank the organizing committees of the various logic programming conferences, symposia, and workshops who made it possible to collaborate in interesting places when it was too inconvenient or tedious to drive 20 miles across Vancouver.

Finally, let us state we are indebted to too many people to mention individually. Our apologies to those whose names are not explicitly stated here.

Table of Contents

PART 0

INTRODUCTION AND HISTORIC OVERVIEW

The principles of logic programming have been applied to formal and natural language processing since 1972, mainly through the Prolog language. Starting from applications such as question answering systems, many interesting problems in natural language understanding were studied with the new insight that logic is a programming tool – an insight very much in line with previous uses of logic in computational linguistics. Thus, areas such as formal representations of natural language, grammar formalisms, methods for analysis and generation, and even very specific linguistic aspects such as coordination evolved in directions typical of the context of logic and of logic programming. In the formal language area, logic grammars have been used for implementing recognizers and compilers, also with good results, particularly in the conciseness of the systems obtained.

Traditionally, logic was considered one of the most appropriate tools for representing meaning, due to its ability to deal formally with the notion of logical consequence. Representing questions and answers in logical form typically required some extensions to classical predicate calculus. In the recent past, logic was also used to represent the information to be consulted in question-answering systems. Here also, departures from first order logic were necessary, and in general, parsing knowledge, world knowledge and the meaning of questions and answers were represented and handled through quite different formalisms, resulting in the need for interfaces to link them together.

The introduction of Prolog (PROgrammation en LOGique) by Colmerauer and others made it possible to use logic throughout, minimizing interfaces from one formalism to another. World knowledge can be represented in logical form, through facts and rules of inference from which a Prolog processor can make its own deductions as needed. Extracting logical consequences amounts to hypothesizing them and letting Prolog deduce, from the facts and rules stored, whether they indeed are logical consequences, and, if they are, in which particular instances. Questioning and answering reduces to hypothesizing the question's content and letting Prolog extract instances, if any, that make the question true with respect to the world described. Those instances become answers to the question.

Parsing itself can be left to Prolog, by representing it as a deductive process – i.e., grammars can be described as facts and rules of inference, and sentence recognition reduces to hypothesizing that the sentence in question does belong to the language, and letting Prolog prove this assumption.

The Prolog equivalent of grammar symbols, moreover, are logic structures rather than simple identifiers. This means that their arguments can be used to show and build up meaning representation, to enforce syntactic and semantic agreement, etc. In other words, the very nature of Prolog facts and rules allow us to retrieve instances from parsing that can tell us more than mere recognition: logical representation of the sentences recognized, causes of rejection, such as semantic anomaly, etc. Thus Prolog is eminently suitable for linguistic work.

Nevertheless, writing substantial grammars in Prolog that could in practice be used as parsers did require knowledge of the language (i.e., a computer specialist's mediation), and involved caring for details that belong to the nature of Prolog's mechanism rather than to considerations of linguistics or the parsing process proper.

The introduction of metamorphosis grammars by A. Colmerauer in 1975 was the first step in making Prolog a higher level grammar description tool. Although these grammars are basically a syntactic variant of Prolog, they achieve two important improvements with respect to Prolog's syntax:

1. They allow the direct writing of type-0–like rules (in the sense of Chomsky's formal grammar classification); these rules can have more than one symbol on the left hand side.

2. They hide string manipulation concerns from the user.

Prolog grammars can now be thought of as rewriting mechanisms which assemble and manipulate trees rather than as mere procedures to be described in terms of Prolog rules and facts. In present implementations, logic grammars are automatically translated into Prolog, but they still remain a distinct formalism in their own right, and they do make life easier for the non-computer specialist–e.g., for linguists.

In 1975 then, all the pieces were laid out to build a new synthesis in the design and implementation of natural language consultable knowledge systems. A. Colmerauer exemplified in a toy system about family and friendship relationships how these pieces could be put together. One of the authors, V. Dahl, developed the first sizeable applications of this new synthesis: first an expert system for configuring computer systems, together with R. Sambuc, and then a data base system with French and Spanish front ends. Both systems were written entirely in Prolog. Logic was used throughout: as a programming tool, as the means for knowledge representation, as the language for representation of meaning for queries, and (in the form of Prolog's hidden deductive process) as the parsing and data retrieval mechanism.

Not all of these uses involved the same type of logic: for natural language representation, Dahl used a three-valued set-oriented logical system; for knowledge representation, she developed some Prolog extensions such as domains, set handling primitives, etc. Yet other developments were needed to solve problems specific to Prolog, such as dealing with negation and modifying Prolog's strict left-to-right execution strategy in order to provide a more intelligent and efficient behavior.

These extensions, as well as the link between different logic formalisms used in these systems, were also hidden from the user. From an implementation point of view, the fact that logic was used throughout made the linking of different formalisms a much simpler task than in typical data base systems, resulting in a concise formulation. For instance, the three-valued logic mentioned above was implemented through a Prolog definition of how to evaluate its expressions, which took less than a page of code.

These techniques were soon exported to other data base and expert systems consultable in other languages (English, Portugese, etc); the main feature of the application of these techniques was the striking ease with which the transposition was achieved. An English adaptation of this system was used in a key paper by F. Pereira and D. Warren—"Definite Clause Grammars for Language Analysis." This article analyzed the logic grammar approach and compared it with the Augmented Transition Network approach, concluding that the former approach was superior.

These encouraging results prompted further research: the techniques for language analysis and for modifying execution strategy were adapted into the CHAT-80 system. M. McCord systematized and perfected the notion of slots and modifiers that had been used in the earliest analyzers, achieving a more flexible strategy for determining scope; F. Pereira developed the *extraposition grammar* formalism, specifically designed to make left extraposition descriptions easier. Dahl and McCord then joined efforts to produce a metagrammatical treatment of coordination in logic grammars, and developed as a by-product a new type of them called *modifier structure grammars* (MSGs), these are essentially extraposition rules in which the semantic components are modularly separated from the syntactic ones, and for which the building of semantic and syntactic structure, as well as the treatment of quantifier scoping, of coordination, and of the interaction between the two, is automated. Further work on coordination was produced by C. Sedogbo and L. Hirschman. The notion of automatic structure buildup that resulted from Dahl and McCord's work on coordination was isolated by H. Abramson into the *definite clause translation grammar* formalism (DCTGs). In it, natural language processing power is traded for simplicity (e.g., quantifier scoping, coordination and extraposition are no longer automated), but for other applications, semantic structure buildup is usually enough. The separation between syntactic and semantic rules is also mentioned.

In 1981, V. Dahl generalized extraposition grammars into a more powerful formalism, called *discontinuous grammars*,[1] that can deal with multiple phenomena involving discontinuity: left and right extraposition, free word order, more concise descriptions, etc. Implementation issues were investigated jointly by the authors, by Dahl and McCord, and by F. Popowich. A constrained version of these grammars was investigated by Dahl and Saint-Dizier.

A more interesting subclass of the *discontinuous grammar* family was developed by Dahl and investigated within her research group for the purpose of sentence generation using Chomsky's Government and Binding theory: *static discontinuity grammars* (SDGs). In this subclass, movement phenomena can be described statically, and the power of type-0 rules coexists with the representational simplicity of context-free–like rules (i.e., trees rather than graphs can depict a sentence's

---

[1] The early publications use the term *gapping* instead of *discontinuous*. This name was changed in order to avoid evoking the wrong associations, since the linguistic notion of *gap* is a different one.

derivation). Hierarchical relationships, crucial to linguistic constraints in general, are thus not lost, and linguistic theories can be accommodated more readily, by expressing these constraints in terms of node domination relationships in the parse tree.

Bottom-up parsing has been investigated by A. Porto et al., Y. Matsumoto et al. and K. Uehara et al. Miyoshi and Furukawa have developed another logic programming language specifically suited for object oriented parsing. M. Filgueiras has studied the use of cooperating rewrite processes for language analysis.

The field is active and promising. This book intends both to introduce the main concepts involving language processing developments in Prolog, and to discuss the problems typically encountered and some of the alternatives for solving them.

After an in-depth presentation of the basic material, we provide a wide rather than deep coverage of many of the related topics. Some bibliographic references are mentioned within the text at places where they are directly relevant, and at the end of each part we complete the picture with other relevant bibliographic comments.

Some chapters were written by one of the authors and revised, with suggestions and comments by the other one: chapters 1 – 4, 6 – 8, 10; section 2 of chapter 11; appendix I; and sections 1 and 2 of appendix II were written by V. Dahl. Chapters 5 and 9; section 1 of chapter 11; chapters 12 and 13; and section 3 of apppendix II were written by H. Abramson. Bibliographic commentaries were written jointly.

PART I

GRAMMARS FOR FORMAL LANGUAGES AND LINGUISTIC RESEARCH

Chapter 1

What Are Logic Grammars?

1. Logic Grammars – Basic Features

Logic grammars can be thought of as ordinary grammars in the sense of formal language theory, in that they comprise generalized type-0 rewriting rules–rules of the form : "rewrite α into β," noted:

$$\alpha \rightarrow \beta$$

where α and β are strings of terminals and nonterminals. A terminal indicates a word in the input sequence. A sequence of terminals takes the form of a Prolog list. Nonterminals indicate constituents. In this text, they take the form of a Prolog structure, where the functor names the category of the constituent and the arguments give information like number class, meaning etc.

Logic grammars differ from traditional grammars in four important respects:

1. The form of grammar symbols, which may include arguments representing trees

2. The use of variables, involving unification

3. The possibility of including tests in a rule

4. The existence of processors based on specialized theorem - provers that endow the rules with a procedural meaning by which they become parsers or synthesizers as well as descriptors for a language (e.g., Prolog and its metalevel extensions)

2. Grammar Symbols

Logic grammar symbols, whether terminal or nonterminal, may include arguments (as opposed to the formal grammars of Chomsky's hierarchy). One of the uses of these arguments is to construct tree structures in the course of parsing. A tree such as

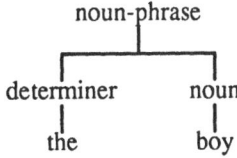

is represented in functional notation:

    noun-phrase(determiner(the),noun(boy))

More generally, arguments have the form:

    root($t_1, \cdots, t_n$)

where *root* is an n-ary function symbol (in our example, the binary symbol *noun–phrase*), and the t_is are arguments, which in turn represent trees. Note that *argument* is used recursively. By convention (as in Prolog) arguments written in lower case are *constants*. An argument can also be a *variable*, in which case it stands for a particular but unidentified tree or constant. Variable names start with a capital. When $n=0$ (i.e., when the root has no branches), the argument is a constant (e.g. "the," "boy") or a tree consisting of just the root. Here are two more sample trees and their functional representations: (Terminal symbols are noted in square brackets.)

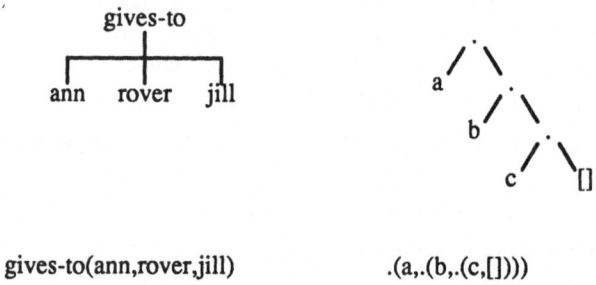

      gives-to(ann,rover,jill)   .(a,.(b,.(c,[])))

In the second of these, the root is the symbol ".", usually used to denote "concatenation." Trees constructed with this binary symbol, as above, are called *lists* and, in logic grammar are interpreted as a string of terminals. They have the simpler equivalent notation:

    [a,b,c]

Summarizing, a logic grammar symbol has the form:

    name($t_1, \cdots, t_n$)

where the arguments t_i are either constants, variables, or trees in functional notation. Terminal symbols are enclosed in square brackets to distinguish them from nonterminal ones.

Here are some (unrelated) sample logic grammar rules:

      1. verb(regarder) --> [look].
      2. verb(third,singular,regarder) --> [looks].
      3. a,[b] --> [b],a.

4. sentence(loves(X,Y)) --> name(X),[loves],name(Y).

We can think of rule 1 as relating *look* both to a verbal syntactic category and to an alternative (French) representation. It could belong to a grammar describing English–French translation. Rule 2 includes some syntactic information about *looks* together with the translation. Rule 3 shows the rewriting of two symbols into an alternative reordering. Rule 4 shows a tree

$$
\begin{array}{c}
\text{loves} \\
\diagup \quad \diagdown \\
X \qquad Y
\end{array}
$$

associated with the sequence: "a name *X*, the word *loves* and another name *Y*." The next subsection shows how rules such as (4) can be applied.

Two of the possible uses of arguments are illustrated here: they may carry syntactic information (such as "third person singular"), or hold some kind of representation for the strings described by the grammar (translations, analysis structures, logical forms, lambda-calculus representations, etc.).

3. Use of Variables and Unification

Variables can be used to represent trees or subtrees whose values are not yet known. For instance, the tree:

$$
\begin{array}{c}
\text{question} \\
| \\
F
\end{array}
$$

containing the variable *F*, stands for the infinitely many trees obtained by replacing *F* by some particular tree, which may in turn contain variables. Here is a logic grammar which we shall call G, that manipulates trees such as the above:

(G1) sentence(question(F)) --> adjective(F),[?].
(G2) sentence(assertion(F)) --> adjective(F),[.].
(G3) adjective(busy) --> [busy].
(G4) adjective(tired) --> [tired].

Although a tree such as *question (F)* stands for an infinite number of trees, in the context of a specific grammar its possible values are defined by the grammar rules. For grammar G above, the possible values are: *question(busy)* and *question(tired)*, corresponding respectively to the sentences: *busy?* and *tired?*. We next develop this correspondence.

3.1. Unification

The fact that grammar symbols may have arguments which may in turn include variables (i.e., represent trees that are partially or totally unknown), means that rule application must involve more than simple replacement. If for instance, we are given the string adjective(X),[?]

and asked to apply a rule to it, we can only apply, say, rule G3, once we have assigned the value *busy* to variable X. Applying the rule would then yield the string:

[busy],[?]

which is equivalent to [*busy,?*] by virtue of the notation for lists. More generally, rule application involves trees. Of need, assignment becomes a more general concept called *unification*. Given two trees, which may contain variables, unification is the process of finding values for those variables, if such values exist, that will make the two trees identical. The set of value assignments {variable = value} that makes two trees equal is called a *substitution*. For instance, the trees

```
      f                     and                    f
     / \                                          / \
    a   X                                        a   b
```

unify with the substitution {X=b}. After unification, both trees have the form

```
      f
     / \
    a   b
```

The trees

```
        f                  and                    f
       / \                                        / \
      X   g                                      b   g
         / \                                        / \
        a   X                                      Y   Z
```

unify into

```
        f
       /\
      b  g
        / \
       a   b
```

with {X=b, Y=a, Z=b}[1]

When the trees to be unified share any variable names, all occurrences of these must be renamed in one of the trees before attempting unification. For instance, from the trees:

---

[1] Z's value is *b* because it unifies to *X*, which in turn has the value *b*.

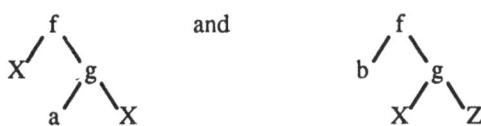

we should first obtain two trees with no variable names in common, by, for instance, renaming X in the second tree into a new variable name Y (or, alternatively, renaming both X's in the first tree into Y).

Prolog automatically renames variables as needed, so we shall assume this task to be done in the remainder of the work. As a corollary to this, we can use the same variable names over and again in different rules without danger of interference. Only when a variable name is used more than once in one and the same grammar rule, does it refer to the same entity.

3.2. Derivation Graphs

We can read logic grammar rules in two basic different ways: declaratively and procedurally. In the *declarative* reading, we view rules as definitions, e.g.:

> (G1) A sentence represented *question (F)* is an adjective
>   represented F followed by a question mark.
> (G3) An adjective represented *busy* is the word *busy*.

In the *procedural* reading, we view each rule as a rewriting specification for a parser. In what follows we assume a top-down left-to-right strategy (as in Prolog), but it should be borne in mind that the logic grammars, viewed as formalisms, are independent of their particular implementations.

We can represent the parsing process by a *derivation graph*, which depicts the history of rule applications needed. For instance, for *busy?* we would have the graph:

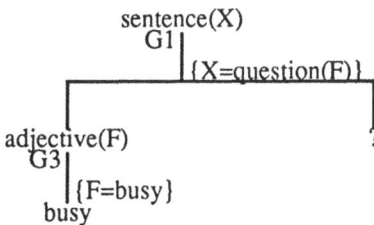

A derivation graph starts with a nonterminal symbol called the *starting symbol*, and grows by rule application. At each point in the derivation, given the string of symbols

$$\alpha\beta_1 \cdots \beta_n\gamma$$

where the substring $\beta_1 \cdots \beta_n$ can be unified with the left hand side of a rule $\beta_1' \cdots \beta_n' \longrightarrow \delta_1 \cdots \delta_m$ with substitution θ , we expand the graph (i.e., apply the rule) by drawing:

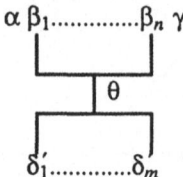

where $\delta'_1 \cdots \delta'_m$ are $\delta_1 \cdots \delta_m$ affected by the substitution θ. We label each arc corresponding to a rule application by the substitution used. We may also add the rule number in the label. The new string is now $\alpha\delta'_1 \cdots \delta'_m\gamma$, for which the process can be repeated, until a desired string of terminals is obtained.

When we draw a derivation graph by hand, we are careful about choosing the appropriate rules when there is a choice. In the above graph, we could have chosen G2 to expand *sentence (X)*, but such a choice would not have led us to derive our target sentence *busy?* Prolog achieves the effect of choosing the appropriate rules through backtracking upon choices made once they prove wrong.

3.3. Symbol Arguments: Producers and Consumers of Structure

Here is another derivation graph, this time for the following grammar, which we call H, with starting symbol *sentence (X)*.

(H1) sentence(S) --> proper_noun(K),verb(K,S).
(H2) proper_noun(ann) --> [ana].[2]
(H3) proper_noun(tom) --> [tomas].
(H4) verb(K,laugh(K)) --> [rie].

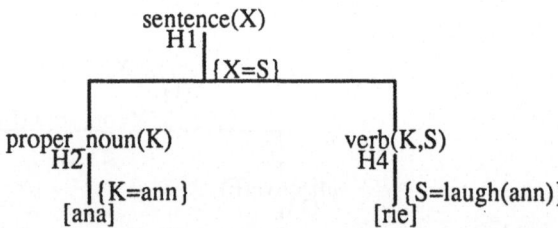

Notice that once a variable takes a value, this value is propagated to each of its occurrences in the graph. Thus, when applying rule H4, we use the known value of *K=ann*. Upon completing this derivation, *X* has therefore the value: *laugh(ann)*, the grammar's intended representation for *ana rie*. By changing the

---

[2] Because variables start with a capital, we must write (constant) proper names in lowercase letters, unlike in ordinary English.

value of the arguments in the grammar we can obtain alternative representations of terminal strings.

In procedural terms, grammar symbols can be viewed as producers and consumers of structure: the *proper-noun* symbol above produces the value *ann,* which is then consumed by *verb* in order to produce the structure *laugh (ann).* (We are assuming a left-to-right order in the derivation.)

Derivation graphs are sometimes useful to visualize how a given grammar relates to given sentences. Whether one thinks of rules declaratively or procedurally is mostly a matter of personal inclination, but it must be noted that the fact that we do not *need* to think of them procedurally is a major high level feature of logic grammars. Their beauty is that the Prolog processor will *draw* the derivation graph automatically and exhibit the substitutions obtained. For instance, if, given the four rules in grammar H, we now enter the query:

> ?-parse(sentence(X),[ana,rie]).

Prolog will print

> X = laugh(ann)

in reply. The symbols "?-" are Prolog's prompt for a query. Notice that grammar H is more than a mere recognizer: it can build up structural descriptions in another language. To obtain them in English, we must simply replace the Spanish words by their English equivalents in the grammar. Omitting the arguments would yield just a recognizer or generator.

4. Tests within Rules

In transformational grammars, some rewriting rules make use of conditions. Unification can be used not only to build structure representing a sentence, but also as an implicit way of specifying tests, the equivalent of the condition. For instance, if we want to make sure that *rie* (laughs) will never be associated with a nonhuman subject, we may add semantic type information in the arguments, and force type agreements through unification. We rewrite grammar H as follows:

> (H1') sentence(S) --> proper_noun(T,K), verb(T,K,S).

> (H2') proper_noun(human,ann) --> [ana].
> (H3') proper_noun(bird,tom) --> [tomás].

> (H4') verb(human,N,laugh(N)) --> [rie].

Notice that the derivation graph for "tomás rie" is now blocked, as the types *bird* and *human* cannot be unified. Another way of specifying tests is through *procedure calls.*

The right-hand side of a logic grammar rule may contain procedure calls as well as grammar symbols. A procedure call has the form

> $\{P_1,P_2,\cdots,P_n\}$ with $n \geq 1$

where the P_i are predicates for which Prolog definitions have been given.

For instance, let us assume a Prolog definition for the predicate *human (X)*, (e.g., one that states that Rousseau and Gauguin are human). Then we can replace rule H4 in the previous section by a rule that checks semantic agreement as well, by requiring the verb's subject to be human:

  (H4)verb(K,laugh(K)) --> [rie],{human(K)}.

This rule can be read: *rewrite a substring of the form*

    verb(X,Y)

into rie, provided that *Y can be unified with laugh (X),* and that *X* satisfies the *human* property.''

Notice that if we specify this test through unification, no extra rules are needed: we simply add type information to the existing rules. The procedure call option, on the other hand, assumes a Prolog definition for the *human* predicate, e.g.:

    human(rousseau).
    human(gauguin).

We shall not discuss the details of Prolog definitions or calls until later, so as not to develop two different notations until we've had more practice with logic grammars. Let us merely mention that the two formalisms (Prolog and logic grammars) are basically syntactic variants of each other.

5. Operators

In the grammar H' in the previous section we can reduce the number of arguments if we combine the type and the representation of a name into one argument. We can do this by using a binary function symbol (say, f) and writing $f(T,K)$ into one argument rather than having the two arguments T, K. Grammar H' becomes:

  (H1'') sentence(S) --> proper_noun(X), verb(X, S).

  (H2'') proper_noun(f(human,ann)) --> [ana].
  (H3'') proper_noun(f(bird,tom)) --> [tomás].

  (H4'') verb(f(human,N),laugh(N)) --> [rie].

Rule (H1'') is now more compact and readable than H1'. In this small example, it might not appear as much of a gain, but in large grammars, it is useful to be able to reduce the number of arguments by combining them into a tree. Since the functional representation for trees, however, is somewhat cumbersome to read, we can make use of a Prolog feature that allows us to write functional expressions in infix, suffix or prefix notation. Thus, if we declare, say, "-" to be a binary operator in infix notation, Prolog will know that an expression t-k stands for the functional expression -(t,k); graphically for the tree

Any book on Prolog explains how to declare Prolog operators. Assuming "-" has been declared, we can write the grammar H" in the easier-to-read form:

(H1''') sentence(S) --> proper_noun(X), verb(X,S).

(H2''') proper_noun(human-ann) --> [ana].
(H3''') proper_noun(bird-tom) --> [tomás].

(H4''') verb(human-N,laugh(N)) --> [rie].

Notice that the operator used is a hyphen, whereas the symbol we use for compound names is an underscore.

Exercises

1. Write a derivation graph for the sentence *ana rie* using grammar H'''.

2. Devise a typing scheme that uses unification for checking semantic agreement, and in which a type's representation shows the inclusion relationships of that type to its supersets. Assume that for any two types, either one is included within the other one, or they are disjoint. For instance, in representing the types *human, employee,* and *salesman,* the representation of *salesman* should somehow show that salesmen are employees, who are in turn human. Thus when parsing the sentence, "Which salesmen live in Vancouver" the representation of the type *human* required by *live* should unify with the representation of *salesmen.* Write a small grammar that implements these types. □

6. Analysis and Generation

As mentioned before, we can think of a logic grammar as a *description*, in declarative terms, of a given language, but also as a set of procedures for analyzing or generating a string in that language. Derivation graphs can be constructed from a grammar, and Prolog does this automatically. In section 3.3 we saw that, with respect to grammar H, we can simply type a query:

?-parse(sentence(X),[ana,rie]).[3]

and Prolog will use the rules in H to *draw* the appropriate derivation graph that will unify X with *laugh(ann)*—in other words, one that will analyze *ana rie* into

---

[3] Appendix I shows a Prolog definition of the predicates *parse* and *generate*; the notions for understanding such definitions are given in chapter 2.

laugh(ann).

But we can also use the same grammar H, unmodified, to synthesize a sentence from its representation. We merely query:

   ?-generate(sentence(laugh(ann)),X).

and as a result Prolog will now print:

   X = [ana,rie].

Thus, sentence parsing and generating procedures are provided by the Prolog processor. If we consider logic grammars *in conjunction with Prolog*—or with any other logic programming system accepting them—rather than as abstract, descriptive formalisms, they become procedures for parsing and generating sentences while maintaining their declarative eloquence.

However, making a more realistic grammar work in both ways requires careful design. Prolog's extra-logical features, to be mentioned later, can prevent a grammar that works one way from working the other. Most of the research to date on logic grammars concerns parsing applications, and very little has been done on generation.

Exercises

1. Write a grammar for generating English sentences of the form: *proper_noun, verb* (e.g., "Ann laughs"), from the logical forms produced from Spanish by grammar H in section 1.3.3 of chapter 1.

2. Query both grammars for translating Spanish into English.

3. Fuse the two grammars into one. (Hint: use a parameter to indicate what language you're dealing with.) □

7. A Formal Language Example

Logic grammars not only serve to describe natural languages, but also artificial ones. This, as will be seen later (see chapters 5, 8, and 11) has important applications, e.g., for compilers. We next compare different logic grammars for the formal language $\{a^n b^n c^n\}$.

Borrowing from formal language theory, we might decide we need more than context-free rules, since this language is a typical example of a context-sensitive one. We might therefore produce the grammar:

   (1) s --> [a] s [b] c.
   (2) s --> [].
   (3) c [b] --> [b] c.
   (4) c --> [c].

Notice that the *c* generated by *s* must in this case be a nonterminal, since otherwise rules 3 and 4 would violate the convention that all rules start with a nonterminal.

Now, because logic grammars allow symbol arguments and procedure calls, we can also choose to express the context-sensitive information through arguments and calls, thus making the grammar look like a context-free grammar.

s --> a(n),b(n),c(n).

a(1) --> [a].
a(N) --> [a],{M is N-1},a(M).

b(1) --> [b].
b(N) --> [b],{M is N-1},b(M).

c(1)
c(N).

We might even collapse the last six rules into two by generalizing even further:

s --> sequence(n,a),sequence(n,b),sequence(n,c).
sequence(1,X) --> [X].
sequence(N,X) --> [X],{M is N-1},sequence(M,X).

Section 2.6 in chapter 8 shows other ways of defining the same grammar.

Exercises

1. Define a logic grammar for recognizing simple arithmetic operations involving addition, subtraction, multiplication, and division.

2. Define a logic grammar for the language $\{a^n\ b^m\ c^n\ d^m\}$. □

Chapter 2

Relation of Logic Grammars to Prolog

This chapter is not meant to provide a thorough coverage of Prolog, but to examine its relationships with logic grammars. For more comprehensive coverage, there are several introductory books. (See Bibliographic Commentary at the end of this part.)

1. Basic Prolog Concepts

We have mentioned that some logic grammars (DCGs, MGs– cf. chapter 6, sections 1 and 2) are basically syntactic variants of Prolog. In Prolog, symbols are noted the same but are called *terms*, and rules have the form:

$$P :- P_1, P_2, \ldots, P_n$$

to be read declaratively as:

"P is true if P_1 and P_2 and $\cdots$ and P_n are all true for all values of the variables in the rule,"

or procedurally, as:

"To execute goal P, execute subgoals $P_1, P_2, \ldots, P_n$."

Notice that the left hand side of a Prolog rule has just one term, whereas in some logic grammars, a rule may contain more than one left-hand side symbol.[1]

Unconditional rules are called *assertions* and noted:

$$P.$$

Queries are noted:

$$? - P_1, P_2, \ldots, P_n$$

They are read declaratively as : "Do there exist values for the variables in the query such that P_1 and P_2 and ... and P_n are all true?" and procedurally, as: "execute the goals $P_1, P_2, \ldots, P_n$."

Executing a goal amounts to drawing a derivation tree (the analogue of a derivation graph in logic grammars). We start with the goal list as the root of the tree, and expand the tree by executing each subgoal in sequence. Executing a goal is analogous to applying a grammar rule: look for a Prolog rule whose head matches the goal, and replace the goal by the rule's right hand side affected by the unifying substitution used. Then we execute in sequence the goals thus obtained,

---

[1] In practice, these rules are usually restated as Prolog rules as we shall see later–extra left-hand side symbols are subsumed into the arguments by the compiler of the grammar rules to Prolog rules.

backtracking upon partial failures, until no more goals remain (success), or no rule applies (failure). For instance, if we have the Prolog rules:

  (1) loves(X,alice):- mad(X),hatter(X).
  (2) mad(Z):- logician(Z).
  (3) hatter(lewis).
  (4) logician(lewis).
  (5) author(lewis).

and the query:

  ?-author(Y),loves(Y,alice).   ("What author loves alice?")

we construct the tree:

$$
\begin{array}{c}
\text{author(Y), loves(Y,alice)} \\
(5) \quad | \quad \{Y=\text{lewis}\} \\
\text{loves(lewis,alice)} \\
(1) \quad | \quad \{X=\text{lewis}\} \\
\text{mad(lewis), hatter(lewis)} \\
(2) \quad | \quad \{Z=\text{lewis}\} \\
\text{logician(lewis), hatter(lewis)} \\
(4) \quad | \\
\text{hatter(lewis)} \\
(3) \quad |
\end{array}
$$

The variable values obtained through the substitutions used constitute an answer to the query.

As in grammars, several rules may apply in trying to execute a goal. Prolog "chooses" a suitable one through trial and error, backtracking to previous points in the tree when it fails and other choices remain untried.[2]

The nondeterminism induced by alternative choices can be controlled, however, through an extra-logical feature: the cut operator (!). Executing '!' has the side effect of suppressing all postponed choices for all goals to be executed, starting from the one which activated the rule containing '!', to the one which precedes '!' in the rule. Other extra-logical features include input-output rules and arithmetic operation rules.

---

[2] Alternative suitable rules will be tried in order, possibly giving rise to alternative parsings.

Some of the grammar types presented in section 2 have been implemented by compiling grammar rules into Prolog rules; others are interpreted through a Prolog program.

2. Prolog as a Language Recognizer

We can use Prolog statements to define what strings a given language is composed of. For instance, for the simplified subset of English containing sentences with only a proper noun and an intransitive verb in simple present tense, third person singular, we can write:

```
sentence([P,V]):- proper_noun(P),verb(V).
proper_noun(julia).
proper_noun(tom).
        .
        .
        .
verb(runs).
verb(plays).
        .
        .
        .
```

Next we can query this program for acceptance of given sentences, e.g.

```
?- sentence([tom, plays]).
?- sentence([julia, runs, away]).
```

Of course, the first of these queries will yield "yes" as a response (the sentence "tom plays" has been recognized), and the second one, "no" ("julia runs away" is not a sentence in the language).

More complex grammar rules, however, often involve lists of words rather than individual words as above. Instead of using a variable for each word explicitly, (as in *[P,V]* above), we can denote *sublists* with variables, and regard each nonterminal as a consumer of part of the input list, and a producer of what is left. For instance, *noun_phrase* would either consume a proper noun or a determiner followed by a noun, and produce the rest of the input string as the remainder to be analyzed:

```
noun_phrase(X,Y):- proper_name(X,Y).
noun_phrase(X,Y):- determiner(X,Z),noun(Z,Y).
proper_name([tom|X],X).
        .
        .
        .
determiner([the|X],X).
        .
        .
        .
noun([clown|X],X).
```

.
.
.

In analyzing "tom runs," *noun_phrase* would receive the whole input sentence as
the value for X, and *proper_name* would consume *tom* from the front of that
string, leaving the remainder $\overline{Y}=[runs]$. For "the clown runs," the second rule for
noun_phrase would unify Z to $[clown,runs]$ and Y to $[runs]$. Thus *noun_phrase*
has consumed two words from X.

The following rules complete the above mini-grammar:

    sentence(X,Y):- noun_phrase(X,Z),verb(Z,Y).

    verb([runs|X],X).
    .
    .
    .

Thus, each nonterminal N can be written as a predication

    N(I,O)

where I stands for the list of input words, and O for the sublist of I that remains
after consuming some of the front words while analysing them as an N.

Queries now look like:

    ?-sentence([the, clown, runs],[]).
    ?-sentence([tom, plays],[]).

The empty list in the second argument position indicates that the list of words in
the first argument position forms a sentence only if all the words are consumed by
the *sentence* predicate.

Of course, had we used a logic grammar instead of plain Prolog, we would not
have had to worry about handling the pieces of input string explicitly. Logic
grammar formalisms hide this string manipulation from the user, in general, by
means of a (also Prolog written) compiler that automatically transforms logic
grammar rules into Prolog statements, by adding the input and output string arguments to the arguments that each grammar symbol already has.

3. Prolog as a Structure Builder

Meaning representations for the sentence being parsed are constructed in the same
way as in logic grammars. For instance, our first grammar in section 3.3 of
chapter 1 can be modified to build a logical representation in French:

    sentence([P,V],R):- proper_noun(P,R1),verb(V,R1,R).
    proper_noun(julia,julie).
    proper_noun(tom,thonias).
    .
    .
    .

```
verb(runs,K,court(K)).
verb(plays,K,joue(K)).
```

Proper_noun produces a partial structure from P, which the verb then uses, together with V, to produce the final representation.

A sample query would be:

```
?-sentence([julia,plays],R).
```

to which Prolog will respond:

```
R=joue(julie)
```

Section 2.2 shows how, in the same way as string manipulation can be made invisible to the user, structure buildup can be largely automated.

The relationship between Prolog and logic grammars, therefore, is in essence that of one being a syntactic variant of the other. But only for the very basic types of logic grammars is this correspondence direct (e.g. when the main difference is basically in the hiding of string manipulation arguments). As we move on to more evolved logic-based grammars, however, the degree of syntactic liberty and of automation of cumbersome tasks grows enough to make them special formalisms in their own right. Besides, we can think of them conceptually, independently of any particular language we might choose to implement them in. The parsers might not even be logic-programming based, but still, viewed as grammar formalisms, they would have a conceptual existence quite distinct from Prolog or from logic programming.

Exercises

1. Assume you represent natural language sentences by graphs in which the labels correspond to each word, and the boundaries between words are named, e.g.,

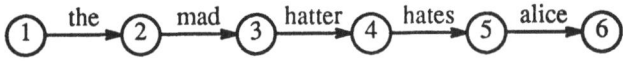

Sentences can then be input through lexical definitions as in:

```
determiner(the, 1, 2).
```

(there is a determiner labeled "the" stretching between points 1 and 2).

a. Use the lexical definitions to describe the more structured constituents, e.g.,

```
noun_phrase(X,Y) -->
        det(_,X,Z),adj(_,Z,W),noun(_,W,Y).
```

In the nonterminal symbols the "_" stand for so-called anonymous variables, variables that we will not and cannot further refer to. Notice that these correspond, graphically, to drawing successive arcs during the analysis of the sentence, e.g.,

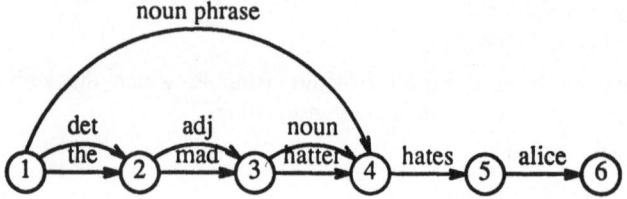

Check the correctness of your grammar by drawing a derivation graph.

b. Change the names of the boundaries so that if a word *w* labels an arc

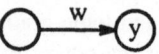

then the starting point is named [wly]. The last point is represented by []. What does your grammar look like now? How do you query it? How does this relate to the idea of symbols as producers and consumers, and to the querying convention shown in chapter 1, section 3.3? □

Bibliographic Commentary for Part I

The first formulation of the parsing problem in terms of Horn clause logic was developed by A. Colmerauer and R. Kowalski. Colmerauer and his group, working from his Q-Systems , developed the first version of Prolog as a tool for natural language processing (Colmerauer 1973; Roussel 1975); however, it was soon clear that Prolog had many general AI applications other than just language processing. Colmerauer then went on to develop Metamorphosis Grammars, an extension of Prolog specifically designed for language processing. They were described in Colmerauer (1978) based on a 1975 technical report. D. Warren and F. Pereira restricted this formalism into definite clause grammars (DCGs) and made them widely known through their article (Warren and Pereira 1980). Applications to natural language processing then began to mushroom. The ones mentioned in Part 0 are covered in the publications: Dahl (1976, 1977, 1979, 1980, 1981, 1982, 1984a,b), Dahl and Sambuc (1976), Filgueiras (1986), Hirschman (1987), Warren (1981, 1982), McCord (1982), Dahl and McCord (1983), Pereira (1981), Dahl and Abramson (1984), Abramson (1984a), and Sedogbo (1985).

Introductions to logic programming and/or Prolog may be found in Clocksin and Mellish (1981), Sterling and Shapiro (1986), Bratko (1986), Hogger (1984), and Giannessini et al. (1986). Logic as a computational formalism and its applications to many artificial intelligence problems are covered in Kowalski (1979). ''The mechanization of deductive reasoning'' is the subtitle of Robinson (1979), which treats first-order logic and its machine realization by means of resolution. The theoretical foundations of logic programming are thoroughly covered in Lloyd (1987). Clark and Tärnlund (1982) contains many important early papers on logic programming. Recent developments in logic programming may be found in the proceedings of the international conferences as well as in the IEEE symposia. Dahl and St. Dizier (1985, 1987) contain collections of papers from two workshops on logic programming and natural language processing. Pereira and Shieber (1987) provides a good overview of Prolog applications to natural language analysis.

There are now quite a few different versions and implementations of Prolog available both from academic centers and commercial enterprises. These include: C-Prolog from Edinburgh (Pereira 1984); Prolog II and Prolog III from Colmerauer's group in Marseille (Giannessini et al. 1986); MU-Prolog and NU-Prolog from the University of Melbourne (Naish 1982); (Thom and Zobel 1986); and the Prolog compiler known as Sicstus has recently become available from the Swedish Institute of Computer Science (Carlsson 1987). A number of commercial Prolog compilers are available. There are also quite a few implementations of Prolog that can be used on home computers (Clark and McCabe 1984). No endorsements are made of those we have mentioned, and apologies to those we have neglected to mention!

PART II

GETTING STARTED:

HOW TO WRITE A SIMPLE LOGIC GRAMMAR

Chapter 3

Step-by-Step Development of a Natural Language Analyser

We now develop a small parsing grammar, step by step. Although oversimplified, it illustrates practically all of the techniques. Most of the grammar samples are for elementary Spanish. The reader is urged to develop his/her own English or other equivalents to gain further practice and insight.

1. Propositional Statements
We shall develop a grammar for describing simple statements constructed around names, verbs and adjectives. This grammar should accept sentences such as:

> John lives in Cordoba.
> Tom is angry with Joan.
> Mary laughs.
> Ann gives Rover to Jill.

Our first thought might be to write rules such as:

> statement --> name, verb.
> statement --> name, verb, object.
> statement --> name, verb, direct_object, indirect object.
> statement --> name, verb, place_complement.
> statement --> name, verb, time_complement.
>   .
>   .
>   .
> etc.

But these rules are adhoc and redundant. They moreover call for more redundant rules, since we next have to state, for instance:

> indirect_object --> preposition, name.
> time_complement --> preposition, name.
> place_complement --> preposition, name.
> etc.

We can greatly simplify our grammar through generalizing some of its concepts.

For instance, all complements of the verb can be treated through a single symbol *comps,* if we simply state for each verb which are its requirements regarding complements. This can be easily described in the lexicon, e.g.:

    verb([]) --> [laughs].
    verb([arg(in)]) --> [lives].
    verb([arg(dir), arg(to)]) --> [gives].

We have added one argument to the *verb* symbol, indicating what expected complements the verb has. The first rule declares no complements for *laughs* (an empty list of complements); the second one forsees one complement introduced by the preposition *in.* The third rule defines two complements, one introduced directly rather than through a preposition, and the other one, introduced by the preposition *to* (*arg* stands for *argument*).

Our rules for statements can now collapse to the single rule:

    statement --> name, verb(L), comps(L).

And the following rules complete our grammar's first version:

    name --> [ann].
    name --> [tom].
        .
        .
        .
    comps([]) --> [].
    comps([arg(P)|L]) --> comps(L), preposition(P), name.

    preposition(dir) --> [].
    preposition(in) --> [in].
        .
        .
        .

The two rules for complements can process any list of complements expected by any verb, since they give a recursive definition that exhausts the list.

But they are not only useful for complements of verbs. Complements for adjectives ("angry with Joan") or for nouns ("friend of John") can also be handled through the same two rules. We merely need to extend our technique of keeping track of expected complements to adjectives and nouns as well, e.g.:

    adjective([]) --> [intelligent].
    adjective([arg(with)]) --> [angry].

We can now deal with sentences of the form "Tom is angry with Joan" through adding another rule for verb:

verb(L) --> [is], adjective(L).

From this discussion we can see that our aim in describing a given language through a grammar is to generalize as much as we can in order to use the fewest possible number of rules. This will not only provide the most elegant grammar, but also simplify the debugging process and minimize the opportunity for error. This is also the direction that linguistic research is taking: Government and Binding Theory (Chomsky 1981) has proposed a set of principles designed to do away with rule proliferation.

2. Obtaining Representations for Propositional Statements

We would now like to obtain internal representations for the sentences parsed, as a by-product of recognizing them. The internal representation could serve as the interface to a data base in which case it would have to be in clause-like form. In the parsing of a computer language, the represententation would be a parse tree or code. Here we would like to get representations such as:

    angry_with(tom,joan)   for: Tom is angry with Joan,
    intelligent(joan)   for: Joan is intelligent,
    give(joan,rover,jill)   for: Joan gives Rover to Jill,
    etc.

As usual, we start by giving individual representations to each grammar constituent, and then we tie these representations together when describing another constituent that includes them. For instance:

    statement(S) --> name(K), verb(K,L,S), comps(L).

    verb(K,L,S) --> [is], adjective(K,L,S).
    verb(K1,[arg(in,K2)],lives_in(K1,K2)) --> [lives].

    adjective(K,[],intelligent(K)) --> [intelligent].
    adjective(K1,[arg(with,K2)],angry_with(K1,K2)) --> [angry].

The rules for complements remain the same. The rule for *statement* ties up the subject's representation *K* with the representation for the complements obtained in *L* after parsing them, into the representation *S* obtained by the verb. This representation can be thought of as a structure with slots (the variables) to be filled in as a result of processing these complements.

Our next step will be to isolate the vocabulary some more. We shall replace the last four rules by:

    verb(K,L,S) --> verb1(be), adjective(K,L,S).
    verb(K1,[arg(in,K2)],live_in(K1,K2)) --> verb1(live).

    adjective(K,[],intelligent(K)) --> adj1(intelligent).

adjective(K1,[arg(with,K2)],angry_with(K1,K2))
                    --> adj1(angry).

verb1(live) --> [lives].
verb1(be) --> [is].

adj1(intelligent) --> [intelligent].
adj1(angry) --> [angry].
etc.

This modification might appear to introduce unnecessary rules. But another of our guiding principles will be modularity. For example in the above grammar the meaning is carried at the level of *verb;* information on tenses and verb endings could be carried at the lower level. If we can thus isolate the vocabulary definitions, changes to the grammar (such as changes to the lexicon) become a simple matter. In another instance, we can adapt it to process elementary Spanish statements by modifying only lexical definitions (of course, in the case of more complex syntactic structures, more than just the lexicon will have to be modified). For applications forseeing such adaptations modularity takes precedence over conciseness. Our corresponding grammar for Spanish looks like:

S) statement(S) --> name(K), verb(K,L,S),
                    complements(L).

V1) verb(K,L,S) --> verb1(be), adjective(K,L,S).
V2) verb(K1,[arg(in,K2)],live_in(K1,K2)) -->
       verb1(live).

A1) adjective(K,[],intelligent(K)) -->
                    adj1(intelligent).
A2) adjective(K1,[arg(with,K2)],angry_with(K1,K2)) -->
                    adj1(angry).

C1) complements([]) --> [].
C2) complements([arg(P,K)|L]) -->
            complements(L), preposition(P), proper_noun(K).

Lexicon:
 L1) adj1(intelligent) --> [inteligente].
 L2) adj1(angry) --> [enojado].
 L3) verb1(be) --> [es].
 L4) verb1(be) --> [está].
 L5) verb1(live) --> [vive].
 L6) preposition(in) --> [en].
 L7) preposition(with) --> [con].
 L8) pr_noun(joan) --> [juana].

L9) pr_noun(tom) --> [tomás].
L10) pr_noun(london) --> [londres].
etc.

Notice that the verb "to be" appears now in its two Spanish materializations: "ser," for permanent status denotation, as in "Juan es alto" (*John is tall*); and "estar," for transitory status denotation, as in "Juan está enfermo" (*John is sick*). Isolating these notions in the lexicon makes it thus easier to change languages.

Similarly, modifications to the internal representation obtained can be effected without changing all the rules in the grammar: only those rules concerned should be touched.

Figure 1 shows the derivation graph for "Tomás está enojado con Juana" (Tom is angry with Joan). Most of the substitutions shown concern the internal representation, X. Some nonterminal symbols are abbreviated. From the substitutions shown, we can see that X takes the value *angry_with(tom,joan)*.

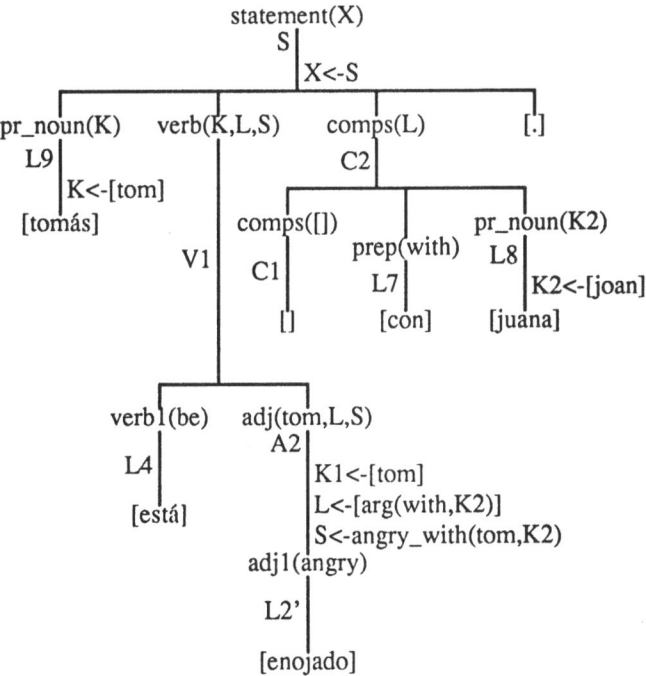

Figure 1.
Derivation for "Tomás está enojado con Juana"

3. Syntactic and Semantic Agreement

Syntactic agreement can be enforced by manipulating features such as gender and number within every rule concerned with syntactic checks. For instance, the modified rules:

> L2) adj1(fem-angry) --> [enojada].
> L2') adj1(mas-angry) --> [enojado].
> L8) proper_noun(fem-joan) --> [juana].
> L9) adj(G1-K1,[arg(with,G2-K2)],angry_with(K1,K2)) -->
>   adj1(G1-angry).

make it impossible to accept a sentence such as:

> Tomás está enojada con Juana.
> *Tom is angry (+fem.) with Joan.*

Semantic constraints can be enforced similarly. For instance, it is well-known that in natural language, there are semantically different kinds of plurals. In "A and B are parallel", for instance, the property of being parallel applies collectively to the set {A, B}, whereas in "A and B are tall" it refers to the elements of the set. Some plurals induce collective relations as the former, and some distributive ones, as the latter. For a referential word (i.e., a noun, verb or adjective) to induce a distributive relation we add a function symbol such as *dr* in the corresponding rule, e.g.:

> A1) adj1(G-K,[],dr(intelligent(K))) --> adj(intelligent).

and establish the convention that unmarked predicates are assumed to be collective. The data base component of our system can thus distinguish each kind of relation and ensure an appropriate interpretation in each case.

Checking and enforcing agreement of types is carried out in a way similar to the treatment of syntactic information. We can represent a type t in a manner that reflects set inclusion relations to other types, e.g.:

> [] & t & t1 & ... &tn[1]

where the ti are types such that $E(t) \subset E(t1) \subset ... \subset E(tn)$, and & is a binary operator in infix notation. For example,

> [] & manager & employee & human

describes a human, which is an employee etc.

Such representations may be *partially* specified, as in

> V & employee & human

which can be matched with all those type representations for types contained in or

---

[1] This is a list-like notation in reverse order, where the nil is fronted and explicit. Unification turns out to work best for our purposes in this ordering, but of course the reverse ordering can also be implemented.

equal to the *employee* type. For instance, V can take the values:

> []
> [] & salesman
> [] & manager

etc., according to the context. In this manner we can check whether somebody is at least an employee.

In general, noun definitions will have the most weight in determining types: since nouns introduce data base domains, their associated types are usually completely specified. Although this convention might result in rejecting as semantically anomalous sentences that might deserve closer inspection (e.g., "Do all the animals speak Latin?"), it would seem a reasonable compromise between speed and coverage.

4. Noun Phrases

When we were dealing with simple noun phrases, consisting only of a name, representing sentences was a relatively simple matter. Basically, name's representations become arguments in a relation, the name of which resembled the head of a phrase (i.e., an adjective, noun, or verb):

> john laughs --> laugh(john)
> john reads ivanhoe to mary --> reads-to(john,ivanhoe,mary)

We now modify our grammar so as to handle quantified noun phrases. Agreement, both syntactic and semantic, is now left out for the sake of clarity.

For the time being, we will only use the traditional logic quantifiers $\forall$ and $\exists$. In section 2.3 of chapter 4 we shall examine subtler representations for natural language quantifiers, since the logical ones cannot adequately render their meaning.

For explanatory purposes, let us imagine a natural language quantifier as a device that creates a variable K and constructs a quantified formula S out of K and two given formulas $S1$ and $S2$ (respectively standing for the noun phrase and the verb phrase's translation, roughly speaking). In terms of grammar rules, this can be expressed as:

> determiner(K,S1,S2,S) --> [det].

where det is a given natural language determiner. Three sample rules follow:

D1) determiner(K,S1,S2, $\exists$*(K,and(S1,S2)))* --> *[el]. (singular "the")*
D2) determiner(K,S1,S2, $\forall$(K,implies(S1,S2))) --> [todo]. (every)
D3) determiner(K,S1,S2, $\forall$*(K,implies(S1,not(S2))))* --> *[ningún].*
> *(no)*

A noun in turn, can be imagined as a device that takes the variable created by the quantification and constructs a relation, as in the following example:

NO1) noun(K,friend(K)) --> [amigo].

We can now relate a noun phrase to a verb phrase, through the rules:

N1) noun_phrase(K,S2,S) -->
    determiner(K,S1,S2,S), noun(K,S1).
N2) noun_phrase(K,S,S) --> name(K).

 S) statement(S) -->
    noun_phrase(K,S2,S),verb(K,L,S2),comps(L).

The first and third arguments of *noun_phrase* are produced by the "noun phrase" device (or, more specifically, inside its rule, by *determiner*), whereas the second is consumed and used by *determiner* to build up S.

$S1$ is not needed as an argument of *noun_phrase,* since it is both produced and consumed inside this rule and does not need to be produced outside it.

To see what values $S2$ might take from outside the noun phrase, consider: "tom reads a book." From "reads," after the subject has been parsed, a skeleton representation $S2$ is obtained, $S2 = reads(tom,X)$, and then in processing the verb's complement, $S1$ becomes $book(X)$. In graphic terms:

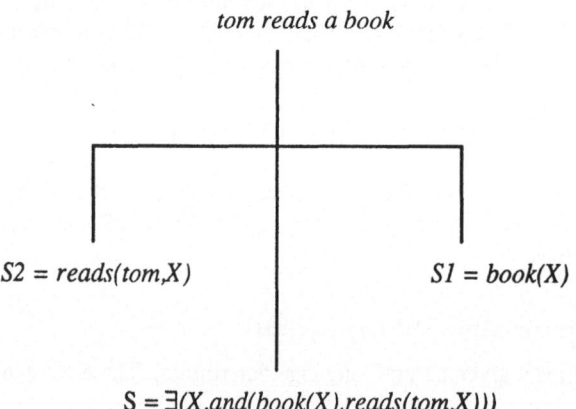

$$S = \exists(X,and(book(X),reads(tom,X)))$$

At this point we urge the reader to draw a derivation graph for the above sentence as an exercise.

Thus, a noun phrase can be regarded as a device taking a formula $S2$ (the verb phrase's representation), and producing a variable K and a formula S that represents the whole statement. In the case of a proper noun, S merely takes the same value as $S2$.

Notice that the order in which these devices are imagined to work is unimportant. They can be regarded as term (i.e., tree) constructors that fill in different slots in those trees they share. For instance, the variable *S2* in rule *S*, which stands for a term of the form

$$r(t_1, \ldots, t_n),$$

which is given such a form by the verb device, while the comps device fills in the values of its arguments. The noun_phrase device, on the other hand, can be considered a consumer of *S2*: it uses *S2* in order to build up *S*. It does not need *S2* to be completely specified, however. It merely fits *S2* into its place in *S*, expecting that sooner or later it will become completely specified.

We can now modify our rules for complements so that they will allow quantified noun phrases as well as proper nouns:

C1) comps([],S,S) --> [].
C2) comps([arg(P,K)|L],S1,S) -->
              comps(L,S1,S2),prep(P),noun_phrase(K,S2,S).

Notice that these two simple rules are enough to handle verb, adjective, and noun complements. All we have to do is modify rules S and N2 as follows:

S) statement(S) --> noun_phrase(K,S2,S),
        verb(K,L,S1),comps(L,S1,S2).
N1) noun_phrase(K,S2,S) -->
        determiner(K,S1,S2,S),
        noun(K,L,S3),comps(L,S3,S1).

and add extra rules for nouns, adjectives, or verbs that accept complements, e.g.:

NO2) noun(K1,[arg(of,K2)],friend_of(K1,K2))
        --> [amigo].

For uniformity, we rewrite NO1 into:

NO1) noun(K,[],friend(K)) --> [amigo].

The compositional buildup of representations for sentences with various complements can be thought of dynamically. In "tom gives a book to each child," for instance, we go from the skeleton representation: *gives(tom,y,z),* to inserting it into another intermediate representation, which is then inserted into the final one. Graphically, we can view it as follows:

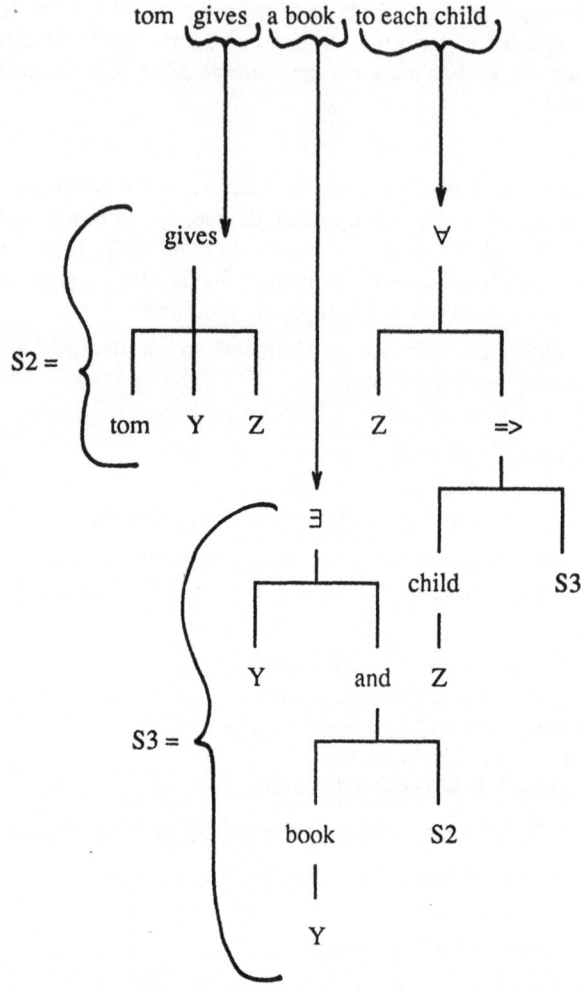

so that the final formula obtained reads:

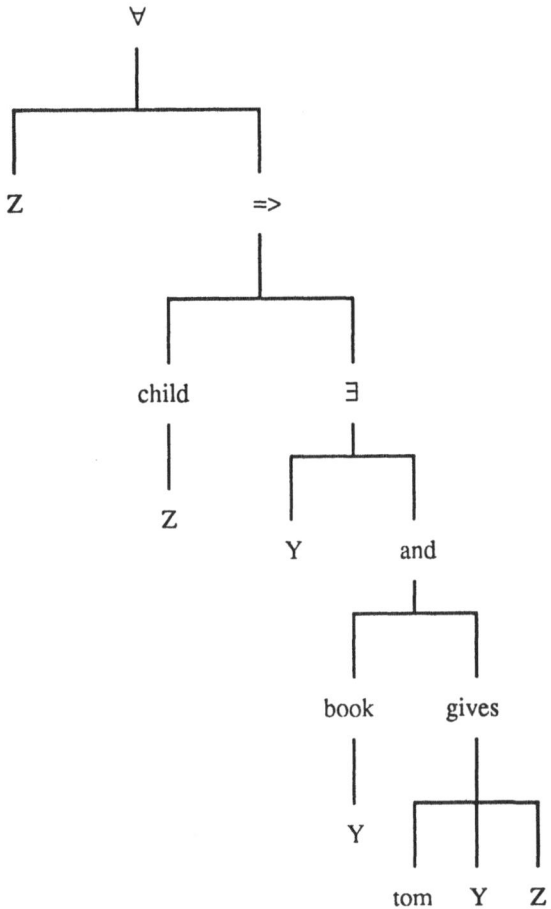

The reader can now make a derivation graph for "El amigo de Juana está enojado con Tomás" (*Joan's friend is angry with Tom*). The internal representation shown in figure 2 should be obtained.

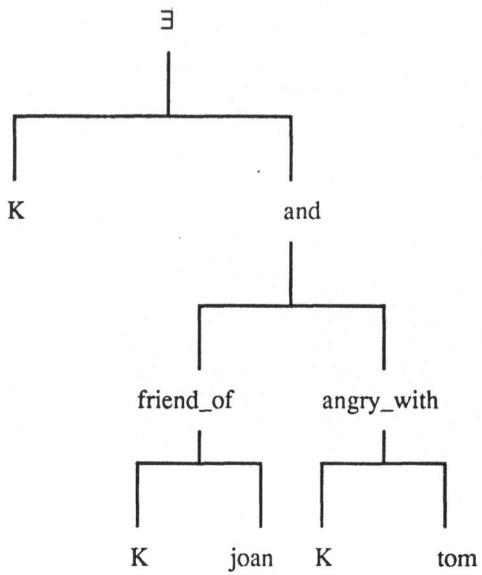

Figure 2.
   Internal representation for
   "El amigo de Juana está enojado con Tomás"

5. Negative Sentences

Semantically, we have already dealt with a form of negation: in section 4, we defined as the internal representation for the determiner "no" the formula $\forall(K,implies(S1,not(S2)))$. An explicit negation in the sentence sometimes corresponds to an explicit negation in the internal representation, but not always. This section examines both cases.

We shall first introduce negation for Spanish, in which the negation "no" (*not*) is simply added before the verb (e.g., "Juan no fuma" — literally, "John not smokes"). Its effect upon the sentence's internal representation should be to negate the formula introduced by the verb phrase, e.g., not(smokes(john)). This can be easily achieved through replacing rule S by:

   S) statement(S) --> noun_phrase(K,S3,S),
                neg(S2,S3), verb(K,L,S1), comps(L,S1,S2).
   G) neg(S,not(S)) --> [no].
   G1) neg(S,S) --> [].

where the neg "device" takes a formula *S* and produces either *S* itself or not(*S*), according to whether the negation particle "no" is absent or present.

It is sometimes useful, when rules become long and difficult to follow, to introduce new grammar symbols that group several other ones by means of an extra rule. This augments the number of rules, but makes them more readable and modular. For instance, we could replace rule S above by the two rules:

    S) statement(S) --> kernel(L,S1,S2,S),
                comps(L,S1,S2).
    K) kernel(L,S1,S2,S) --> noun_phrase(K,S3,S),
                neg(S2,S3),
                verb(K,L,S1).

This grouping into *kernel* is also useful to simplify rules that move verb modifiers, as shall be seen in the next section.

In sentences like "No vino ningún alumno" (*No student arrived*), there is subject-verb inversion, and negation is repeated twice (once in the word "no" (*not*) and once in the meaning of the "ningún" (*no*) quantifier). The internal representation should read: "For every student, it is stated that he did not arrive." If we simply allow the ordinary meaning of "ningún" to be generated, the explicit negation in the sentence will then compromise the correct meaning, by turning the representation:

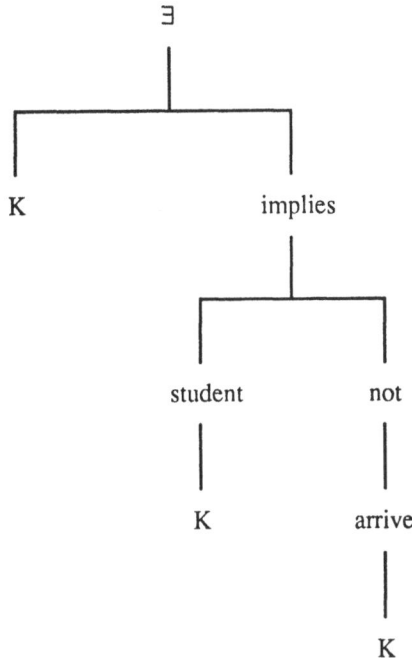

into:

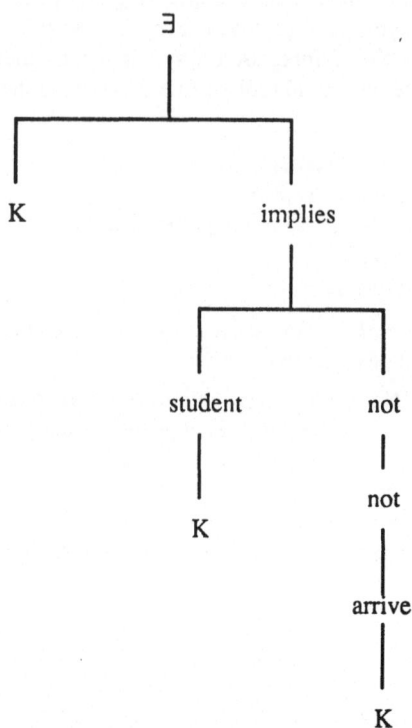

(we are assuming the definition D3 in section 4 for "ningún")

To handle this situation, we take advantage of a nonterminal *case* (*C*) which can explicitly record the role of a given noun phrase as subject. Our rules are augmented as follows:

> K) kernel(L,S1,S2,S) --> modifier(subject-K,S3,S),
>   neg(S2,S3),verb(K,L,S1).
> M) modifier(C-K,S1,S2) -->
>   case(C),noun_phrase(K,S1,S2).

The subject verb inversion rule is as follows. Its application leaves a symbol *inv* as a marker.

> I) modifier(subject-K,S3,S),neg(S2,S3),verb(K,L,S1) -->
>   neg(S2,S3),verb(K,L,S1),inv,modifier(subject-K,S3,S).

This marker is used by rule T below to transform a surface "ningún" into "todo" (every), provided it occurs in an inverted subject. (Left-hand side terminals are synthesized, whereas right-hand side ones are analyzed, so the net effect is that of replacing "ningún" by "todo.") Otherwise, (if rule T does not apply) the markers are erased by rules *M* 1 and *M* 2.

T) inv,case(subject),[todo] --> [ningún].

M1) case(subject) --> [].
M2) inv --> [].

These rules implement our general treatment of negation described earlier.

The nonterminal *case(C)* is important also in handling complement noun phrases. Such noun phrases introduced by a preposition *P* will have associated the case *prep(P)*. Direct object noun phrases will have case *dir,* and so on. We generalize rule C2 to

C2) comps([arg(C,K)|L],S1,S) -->
                comps(L,S1,S2),modifier(C-K,S2,S).

and add

M2) case(prep(P)) --> prep(P).
etc.

Suggested Project

Extend the English grammar developed so far so as to handle negative English sentences (auxiliaries need to be incorporated).

6. Interrogative clauses

As subject-verb inversion has already been defined, we can handle Spanish yes-no questions simply by adding:

SE1) sentence(fact(S)) --> statement(S).
SE2) sentence(yes_no(S)) --> [?],statement(S),[?].

Spanish questions actually begin with a reversed question mark, but most keyboards available to us do not have it.

A yes-no question deals with the truth or falsity of a statement e.g. " Is tom mad, yes or no?" (Notice, by the way, that the analyser actually produces more information than the formula S. It defines whether the statement is a fact or a question. A data base could use this extra information to determine the form of the answer, to identify the set to be retrieved (as in rule SE3 below),etc.).

Wh-questions, on the other hand, often require modifiers to be moved around and replaced by pronouns. For instance, "?Dónde vive Tomás?" (*Where does Tom live?*) can be considered as a variant for "Tomás vive en K" (*Tom lives in K*), in which "en K" has been moved to the beginning of the sentence and replaced by "Dónde."

Relative clauses usually undergo similar transformations. For instance, "El empleado cuya jefa es Juana" (*The employee whose manager is Joan*) can be considered as a variant of "El empleado [la jefa del empleado] es Juana" (*The employee [the manager of the employee] is Joan*), where "del empleado" has shifted to just before "jefa" to be subsumed, together with "la" by the relative pronoun "cuya." To handle these clauses, we use markers in the form of grammar symbols; we move the concerned modifiers and then we use context-sensitive rules to replace the appropriate constituents by a pronoun. We illustrate this for interrogative sentences such as the above example. First we add an interrogative marker:

> SE3) sentence(wh(K,S)) --> [?],wh_1(K),statement(S),[?].

A modifier to be moved can be handled by the extra rule:

> C3) comps([arg(C,K)|L],S1,S) -->
>   moved_mod(K,S2,S),comps(L,S1,S2).

which places it as the first complement. It must now skip the kernel so as to become the head of the sentence:

> SK) wh_1(K),kernel(L,S1,S2,S),moved_mod(K,S3,S4) -->
>   wh_2(K),modifier(K,S3,S4),kernel(L,S1,S2,S).

Finally, it can be replaced by a pronoun:

> PR) wh_2(K),modifier(K,S1,S2) --> [dónde].

Figure 3 shows a simplified derivation graph for "¿Dónde vive Tomás?," from which the internal representation wh(K,live_in(tom,K)) is obtained. Arguments and substitutions are left out in order to emphasize the structure of the derivation. Chapter 5 shows alternative ways of dealing with movement rules.

Suggested Project

Incorporate an English treatment of interrogative clauses into the grammar you have so far developed.

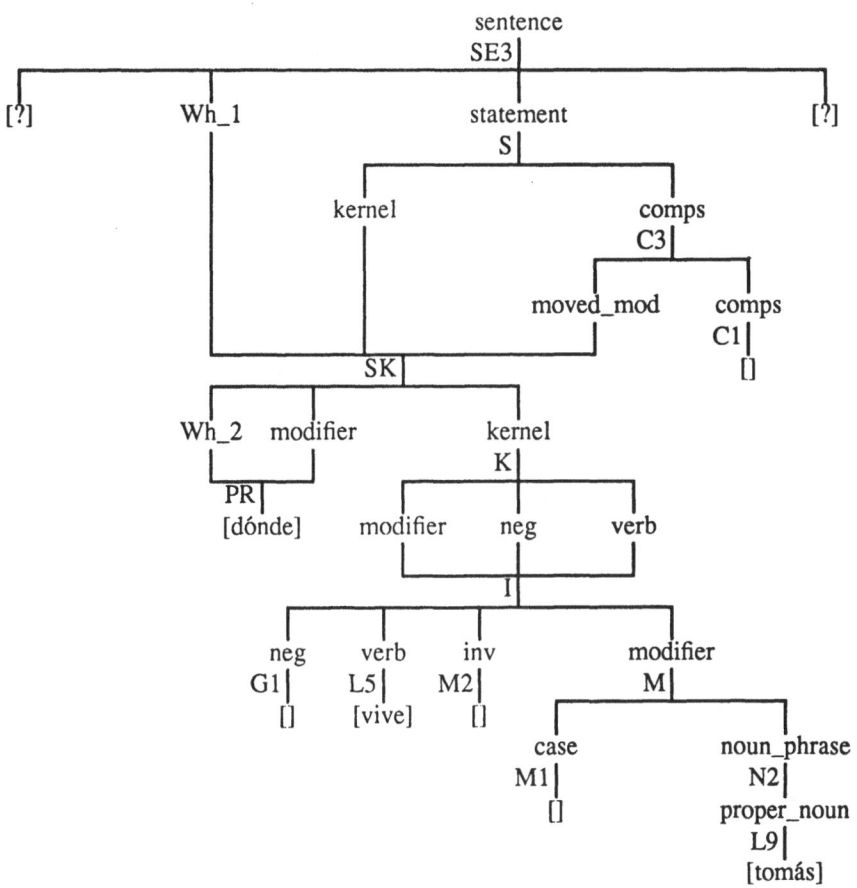

Figure 3.
Skeleton derivation graph for "?Dónde vive Tomás?".

Chapter 4

Choosing Internal Representations for Natural Language

The previous sections have given some of the flavor of how formulas may be produced from sentences through logic grammars. Basically, we have to transform a sentence into a structure in some internal meaning-representation language (first-order logic, lambda-calculus, etc.). The particular language chosen will depend on its suitability to represent the meaning of sentences with respect to the intended use of the structures obtained. We may want to use them for creating, consulting or modifying a data base, for having dialogues with an expert system, for extracting information about a constructive world, as an intermediate form in the translation from one language to another, as an aid for teaching some linguistic theory to a student, etc. Here we restrict ourselves to one of the best studied uses: querying knowledge or information stored.

1. Logical Form for Querying Knowledge Represented in Logic

In a knowledge or information system, such as an expert or a data base system, we want to translate natural language into formal language in order to extract information stored in a computer.

In such systems, logic has lately been playing an important role, particularly in the data-base querying area. Various predicate-calculus based query languages have been developed, especially in connection with E. F. Codd's relational model (1970), owing mainly to the fact that logic is both formalized enough for computer applications and high-level enough to be closer to human comprehension than traditional computer languages.

In 1976 Dahl and Sambuc developed the first Prolog data base/expert system under the name SOLAR16. This system configures a computer around the user's needs.[1] Since then, the use of logic in the form of of logic programming has been extended to *representing* as well as consulting specialized knowledge.

This makes the use of the logical form to represent language even more attractive. If the information stored is represented in logic, it becomes very natural to retrieve it through logic queries. Consider the "human" relationship, expressible through Prolog rules, e.g.:

        human(curie).
        human(gauguin).
        human(rousseau).
        human(X) :- mother(X,Y), human(Y).
        mother(joliot, curie).

The fourth rule states X is human if her/his mother is human.

---

[1] As a first in its field, SOLAR16 had to deal with a number of basic issues. Hence it will be quoted many times. For brevity we will refer to it by name only.

Querying becomes quite straightforward when these definitions are given to Prolog. We merely write:

    ?-human(curie).
    ?-human(X).
    ?-mother(X, curie).
    ?-mother(joliot, X).

to ask such things as: "Is Curie human?", "Who is human?",[2] "Who is Curie the mother of?" and "Who is the mother of Joliot?" Theories such as these can be written by a Prolog programmer, but can also be obtained as the output of such natural language analysers as described earlier.

Ease of expression is due to the fact that Prolog extends our declarative meaning with an operational interpretation, automatically making for us all deductions needed ("drawing" the derivation tree mentioned in chapter 3, section 1. Notice that each argument can take an input or an output role, as determined by the form of the query (e.g., the first argument of "mother" works respectively as output and as input in the last two queries).

From a knowledge representation view, there are several advantages to representing information in this way. These have been discussed in Dahl (1982, 1983) so we merely summarize them here:

1. Deductive capacity (e.g., the inference that Joliot is human is automatically made by Prolog upon, for instance the query: ?-human (joliot)

2. General rules (e.g., the fourth rule above)

3. Recursive rules (ibidem)

4. Inherent modularity (information is naturally broken up into independently meaningful rules)

5. Nondeterminism: alternative answers to the same query are possible (e.g. ?-human(X) results in X successively taking four different values with respect to the above definition)

6. Nondistinction between program and data (specific facts coexist with procedure specifications–Prolog rules–for finding more facts).

Thus, the question of what logic form to use for representing language is affected by our choice of a formalism for representing the knowledge to be consulted through that logic form. If Prolog is used, we are guaranteed automatic handling of queries of the form:

$$?-P_1, \ldots, P_n$$

(i.e., for what values of the variables in the query do P_1 and $P_2 \cdots$ and P_n all hold?).

---

[2] Notice that intensional answers to this question, such as "the mother of a human," are not arrived at by most existing logic programming processors.

2. First-Order Logic Representations

Prolog queries, however, are too restrictive for representing natural language questions. We might consider writing an interface that will translate more evolved queries into Prolog queries. For instance, our computer configurating system SOLAR16 translates French queries into the following well-formed formulas P_i:

1. F
2. exists (X,P1)
3. all (X,P1)
4. not(P1)
5. and(P1,P2)
6. or(P1,P2)
7. implies(P1,P2)

where F is a predicate whose truth values have been defined in the data base and X is a variable.

These formulas are then transformed into Prolog queries through the following Prolog program:[3]

```
tr(exists(X,P)) :- ! , tr(P).
tr(all(X,P)) :- ! , tr(not(exists(X,not(P)))).
tr(and(P1,P2)) :- ! , tr(P1,P2).
tr(or(P1,P2)) :- tr(P1).
tr(or(P1,P2)) :- ! , tr(P2).
tr(not(not(P))) :- ! , tr(P).
tr(not(and(P1,P2))) :- ! , tr(implies(P1,not(P2))).
tr(not(implies(P1,P2))) :- ! , tr(and(P1,not(P2))).
tr(not(all(X,P))) :- ! , (exists(X,not(P))).
tr(not(P)) :- ! , tr(non(P,P)).
tr(implies(P1,P2)) :- ! , tr(or(not(P1),P2)).
tr(P) :- P.
```

Another Prolog program, called the demonstrator, uses these transformation rules to execute the list of atomic goals obtained, in an appropriate order (appropriate with respect to efficiency, given the semantics of the world described in the data base), and also ensures a correct interpretation of negation.[4]

---

[3] A comprehensive explanation of the symbol "!" (cut), a Prolog built-in predicate, can be found in any Prolog manual or book. Here it is simply used to prevent the last clause from unifying P with any of the formulas appearing as arguments of *tr* in the preceeding clauses. Notice that the first "tr" clause for "or" does not have a cut, so that, failing the transformation of P 1, the transformation of P 2 is still available.

[4] When "not" cannot be further reduced, it is transformed into "non" for special treatment by the demonstrator. This treatment extends negation-by-failure (explained in section 2.4 below) to first-order logic formulas: if all variables in a negated formula are quantified inside it, the negation-by-failure of a recursive call to the demonstrator is invoked, else the formula is put in a waiting list.

Most of the above rules are simply applying well-known logical equivalences. The second to last rule transforms a negated formula that cannot be further transformed by any preceeding rule, into a form suitable for correct interpretation of negation as mentioned above.

Although sufficient for this system's needs, this logical form is insufficient for representing many subtleties of natural language. Typically, extensions to standard predicate calculus are needed to represent even modest data base interrogation applications. We next summarize some of the problems involved and examine some possible extensions to solve them.

2.1. Semantic Well-Formedness

Some applications may require the parser to be able to reject semantically anomalous sentences. This can be achieved in two phases, by first producing a purely syntactic representation and then performing all necessary semantic operations and checks on it, or by using a typed predicate calculus, so that checks of meaningfulness are reduced to type checks.

These, as we have shown in chapter 2, can be very economically performed through unification. A more elaborate typing system that takes type inclusion hierarchies into account is presented in Dahl (1981).

2.2. Ambiguity

Some types of ambiguity can also be dealt with through types; specifically, where syntactically acceptable but semantically incorrect readings are concerned. For instance, in: "Give me the ID number of the student that made the highest grade", by requiring the subject of "made" to be of human type, we discard the (syntactically plausible) reading in which the relative clause's antecedent is "the ID number". Different word meanings can also often be choosen according to types. Although very efficient, this method clearly does not address the problem of true (semantic) ambiguity (as in "John saw a man in the park"). Typically, logic grammars will produce all possible analyses through backtracking.

Fernando Pereira (1983) has proposed to eliminate attachment ambiguities in logic grammars by producing a single, normal form where each modifier is attached to the smallest constituent it may modify in the original, ambiguous grammar. Other possible attachments are then derived from the normal form, thus substantially reducing the non-determinacy of the parser. Semantic and pragmatic information that may be used for deriving further attachments are, moreover, separated from the parsing process.

2.3. Natural Language Determiners and Modifiers

Some natural language determiners can be represented using the quantifiers "all" and "exists". For instance, we can represent "No soldier obeyed" by:

    all(X,soldier(X)->not(obeyed(X)))

But how do we represent the subtle meanings in "some", "a", "the", or cardinal numbers?

Other logic representations may render these subtleties appropriately. Here we merely show one of the possibilities.

In Dahl (1981) we describe a treatment of Spanish determiners that relates to the three-branched quantifiers of Colmerauer (1982) and to linguistic research of Hausser (1974). This treatment accounts for those presuppositions induced by natural language quantifiers (e.g., singular "the" presupposes existence and unicity of its referent). We next introduce this quantification treatment.

2.3.1. Three-Branched Quantification

In a first, simplistic approach, we shall represent quantified noun phrases by formulas of the form:

q(K,S1,S2)

where q is the quantifier in question (one of "the", "a", "no", "many", etc.), K is a variable introduced by the quantification, and $S1$ and $S2$ are formulas involving K. If we are dealing with a subject noun phrase, for instance, $S1$ roughly corresponds to the subject's representation and $S2$ to the predicate's representation. Thus, "Every bird flies" is represented:

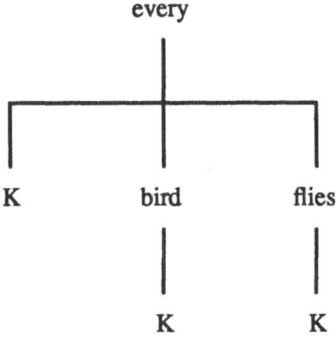

$S1$ and/or $S2$, can, of course, be in turn quantified, as in "Every man loves a woman", which can be represented:

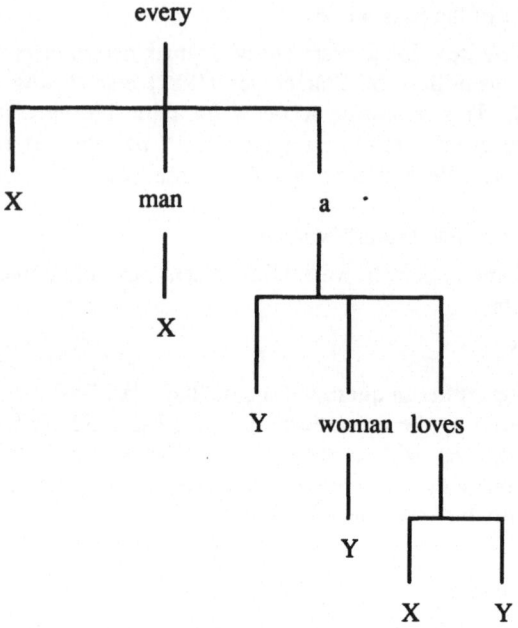

Notice that choosing the right hierarchy of quantifiers is crucial. If we had instead chosen the representation:

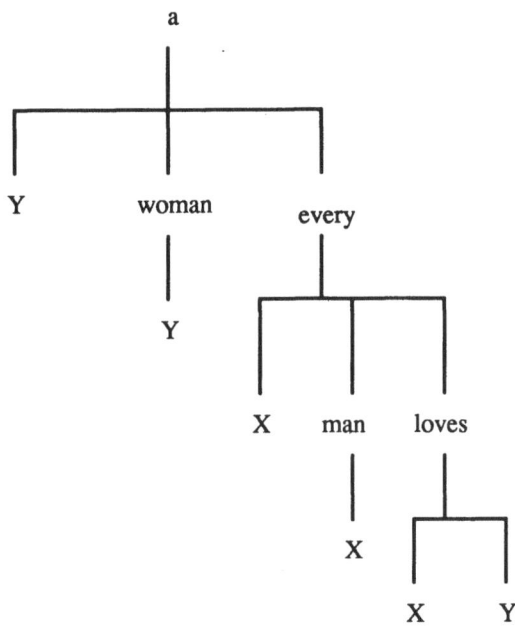

we would be saying that there exists one particular woman (say, Mother-Earth) whom every man loves, whereas in the previous representation we express the more likely meaning: for every man there exists a corresponding (different) woman he loves. Quantification hierarchy is further discussed in section 2.3.2.

Obtaining this representation for our quantifiers in terms of three-branched formulas where the functor is the same as the determiner is very easy. We merely need to change the determiner rules into rules like:

determiner(K,S1,S2,the(K,S1,S2)) --> [the].

determiner(K,S1,S2,a(K,S1,S2)) --> [a].

.
.
.

etc.

But, while easy to construct, these three-branched quantifications do not very accurately express the meaning of each specific quantifier, except in terms of naming the root after it. Many subtler representations are possible; here we present one that proved useful for data base consultation applications. It is based on the idea of paraphrasing sentences in terms of sets and their cardinalities. For instance, "Every bird flies" can be paraphrased as "The set of birds that do not fly is empty." In order to reflect this meaning in our representation, we only need to modify the formula S obtained in the forth argument of "determiner." Instead of $S=every(X,bird(X),fly(X))$, we would have

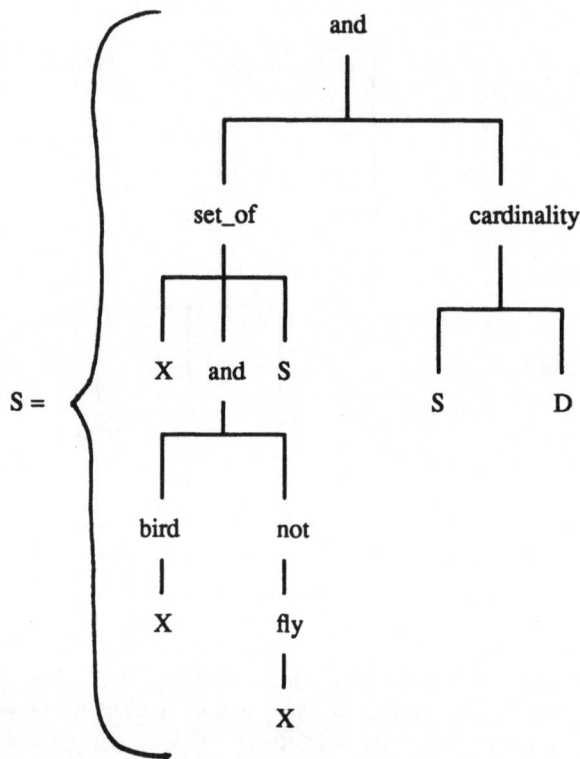

In general, we only need to change a formula:

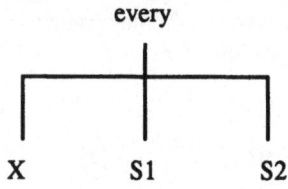

into

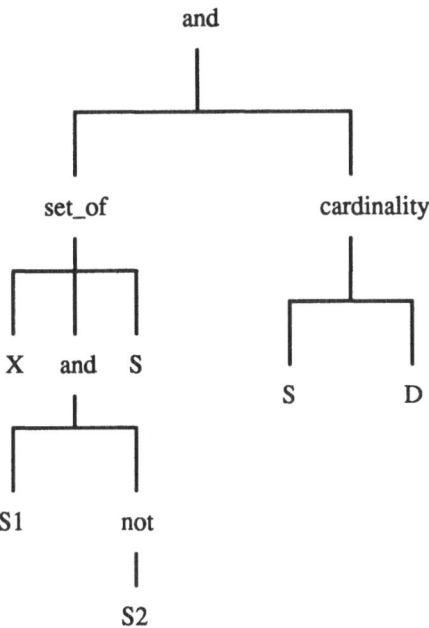

Some determiners implicitly include a presupposition in their meaning. For instance, in "the boy with a red hat" the use of "the" usually tells the hearer that, in the context of this utterance, there is only one boy who has a red hat (otherwise the determiner "a" would have been used instead). Should this presupposition not hold, it would be pointless to talk about "the boy with a red hat." In order to correctly render the meaning of presuppositions, we introduce a formula

   if(P,S)

where P is the presupposition whose satisfaction makes S meaningful.

This representation scheme allows for a semantic detection of failed presuppositions, and is also useful for keeping Prolog's negation-by-failure operator from mischief. But it involves too much element-by-element computation in the calculation of the sets. In SOLAR16 and in CHAT80, another data base query system, dynamic query reordering is used to make negation work reasonably, and presupposition detection is dropped altogether. F. Pique has also contributed work on presuppositions in the logic grammar field.

Since our choice of logical form will be influenced by the processes its formulas will undergo, and since, as we have seen, this processing by a Prolog data base system is one of the obvious possibilities, in the next section we discuss alternative forms for evaluating negated expressions in Prolog.

2.3.2. Quantifier Scoping

Another important representation problem is quantifier scoping in natural language. It is well-known that different combinations of quantifiers in sentences with equal syntactic structures induce different quantifier scopes. For instance, compare the following sentences and their first-order logic representations:

Each woman and each man drove a van.

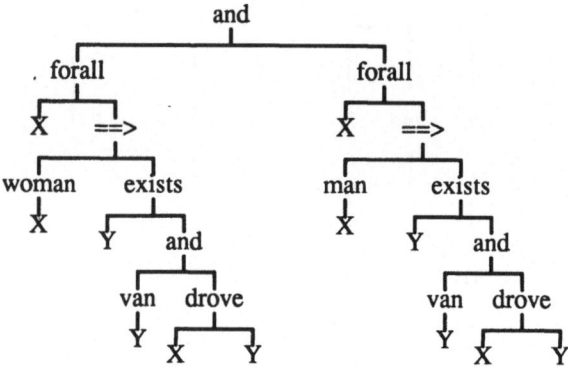

A woman and a man drove each van.

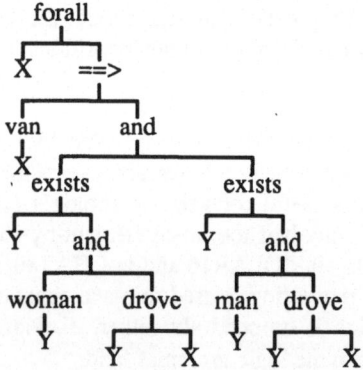

In our Spanish system, we adopted a simplistic but useful strategy, in which the determiner of the subject of a verb dominates all others; the determiner of the complement of a noun dominates that of the noun, and when a referential word (verb, noun or adjective) has more than one complement, quantification takes place from right to left. Although sufficient for our purposes, this strategy is quite inflexible and will result in erroneous representations of certain sentences.

2.4. Negation

When dealing with closed worlds (Reiter 1978), in which a complete knowledge of the domain represented is assumed, we do not need to list all negative information (which is usually very large and often infinite) explicitly. Instead we can establish the convention that all those facts whose truth cannot be proved are assumed to be false.

Prolog's negation-by-failure is defined as follows:

    not(Q):-Q,!,fail.
    not(Q).

If Q can be proved, the cut blocks all other ways of proving *not(Q)* before rule 1 fails. Since rule 2 has been blocked, *not(Q)* fails altogether. Else (if Q cannot be proved) rule 1 is abandoned before the cut has a chance to act, and rule 2 succeeds.

Whenever Q is ground (i.e., contains no variables), such a definition works as expected. But consider what happens for instance if we are given:

    composer(verdi).
    scientist(curie).

and we then query:

    ?-not(composer(X)),scientist(X)

with the intention of finding out who aren't composers but are scientists. Since there is one X that *is* a composer (Verdi), Q will succeed, the cut will be activated, and the whole query will fail.

If instead we query:

    ?-scientist(X),not(composer(X))

Prolog will find the answer: X=curie. As this example shows, within closed worlds, negated predicates are only safe to evaluate when they contain no free variables.

Thus, when choosing a logical form that includes negation, we must ensure, if we are to evaluate it within Prolog's closed world conventions, that any negated formula gets completely instantiated before its evaluation. In the previous section we showed a single quantification mechanism (*set-of(X,T,P,S)*) which introduces typed variables. Evaluating a *set-of* expression involves tying the variable to each domain element in turn before evaluating the formula P in which it intervenes, so if P contains any negations, these will affect only ground subformulas.

The SOLAR16 system, which deals with first-order-logic formulas, gets around this problem with no need for iterating over domains or for having only finite domains, simply through a careful choice of the order in which subformulas are evaluated.

SOLAR16 deals with what we call a *constructive world*, that is, a world that can be viewed as a set of specifications for confining different components into desired configurations or structures. In such worlds, the order in which we execute subtasks is usually crucial. Problems specific to constructive worlds have been discussed by Dahl (1984a).

Briefly stated, SOLAR16 achieves query reordering through a user-invisible coroutining extension to Prolog, implemented in Prolog itself, which examines each successive query and reorders it according to run-time tests.

It assumes numeric delay values associated with each predicate and each state of instantiation of its arguments. For instance, $price(X,Y)$ (i.e., calculate the price Y of a computer configuration X) is assigned an "infinite" (i.e., prohibitively big) delay whenever X is not fully instantiated, and a zero delay otherwise. As this example shows, efficiency and even feasibility are affected by a correct choice of order in evaluating a query, even with respect to ordinary predicates. For $not(P)$, we simply need to associate an infinite delay value if P contains any free variables, and a zero value otherwise.

An interesting variation of this idea was later exploited in CHAT80, where delay values depend upon the size of the relations in the data base, and of the domains over which their arguments range. This gives a useful measure for minimizing search spaces in relational data bases, but other factors must be taken into account where constructive worlds are concerned. For instance, the size of the "price" relation in the SOLAR16 constructive data base seems less important than the fact that it involves a complex tree that might not be fully instantiated.

2.5. Meaning of Plural Forms

Different kinds of plural must be distinguished in our representation formalism. For instance, consider the following sentences and their representations:

Tom and Bill met

$$met(\{Tom, Bill\})$$

Tom and Bill know Latin and Greek

knows(Tom, Latin) &
knows (Bill, Latin) &
knows (Tom, Greek) &
knows (Bill, Greek).

Tom and Bill know Latin and Greek respectively

knows (Tom, Latin) &
knows (Bill, Greek).

In the first sentence, introducing a collective plural, the property of meeting must apply on the whole set of individuals concerned. In the second, this property must be distributed among all possible argument combinations, and in the third, the respective meaning must be rendered appropriately.

Different kinds of plurals can be distinguished in the lexicon, by syntactically marking the relation they translate into. In this respect we can consider relations as typed, and implement different evaluating schemes according to the type involved. We can moreover make use of the third logical value mentioned earlier in order to detect failed presuppositions induced by the particular type of plural involved: notice that both distributive and respective plurals presuppose that the set of formulas to be tested all have the same truth value. Whenever such a presupposition is not satisfied, the plural predication is neither true nor false.

2.6. Representing Sets

It must be clear, from all that precedes, that we need to allow relations to apply on sets. Sets are moreover natural enough in data base applications, as retrieval often concerns sets of objects satisfying certain properties.

Summarizing these points, we need to extend classical first-order logic in order to simultaneously meet natural language processing requirements and information or knowledge handling requirements. In the case of relational data base systems, this led us to define a rigorous, typed, set-oriented, three-valued logical system, which constitutes a less ambiguous formalism than natural language by making some of its semantic features explicit (e.g., distinguishing different kinds of plural in relations, reflecting certain natural language presuppositions, and attaching a semantic type to each term). It has also been carefully designed so as to be safe with respect to Prolog's negation by default. Details on this logical system can be found in (Dahl 1980).

Naturally, many alternative ways of extending first-order logic are possible, and, as long as linguistic research results do not supply us with a comprehensive, all-purpose theory of language, we shall have to continue tailoring our language representation features to measure, with our particular computational application in mind. The fact that we can do so, at least to some useful extent, using logic, and that we can also use logic for representing and querying knowledge, provides an attractive and uniform theoretical setting for further research.

Many interesting problems have been left out of our discusion (coordination, anaphora, etc.), but we hope to have given enough of the flavor in translating language into logic through logic.

3. Lambda-Calculus Representations

Lambda-calculus was introduced by A. Church (1956), in order to have a notation for functions, and for their applications to arguments, which would remove all ambiguity. In this notation, a unary function $f(x)$ is represented as follows, where λ identifies x as being the argument of g:

$$g = \lambda x \cdot f(x)$$

Its application to, say, the value three, is noted:

$$\text{apply} (\lambda x \cdot f(x), 3)$$

and is equivalent to $f(3)$. The application of a function to a value is also called β-reduction.

Because in Prolog substitutions are trivially achieved (through unification), Prolog lends itself quite naturally for β-reduction. In Prolog, we will note the function g

above as: *lambda(x, f(x))*; and its application to the value 3 is noted: *beta(lambda(x, f(x), 3, P)* where *P* stands for the result of the application. Applying an expression *lambda(X,P)* to another expression Q can be simply done by unifying X with Q (provided there is no clash of variables — otherwise we assume previous variable renaming). The result will be (instantiated) P, since unification of course affects all occurrences of X in P. This can be expressed through the following Prolog rule:

    beta (lambda(X,P),X,P).

A Prolog call such as:

    ?-beta (lambda(X,sin(X)),Y,Q)

which can be depicted as

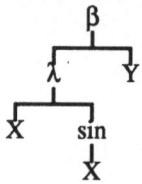

has the effect of unifying X with Y, and Q with sin(Y). We now describe a toy grammar for sentences of the form: "No person has every defect", using procedure calls to the "apply" predicate defined above.

Lambda-calculus is attractive if one views the meaning of syntactic categories as mappings from either variables or other properties into new properties. For example, we can consider the category "name" as a "meaning device" that takes a variable X and constructs a property from it - such as $person(x)$ or $defect(x)$. A determiner, on the other hand, can be viewed as a device that takes two properties (roughly representing the subject and the predicate of the sentence) and constructs a new property tying those two up and rendering the meaning of the specific determiner.

The vocabulary can be depicted as follows, in terms of lambda-calculus:

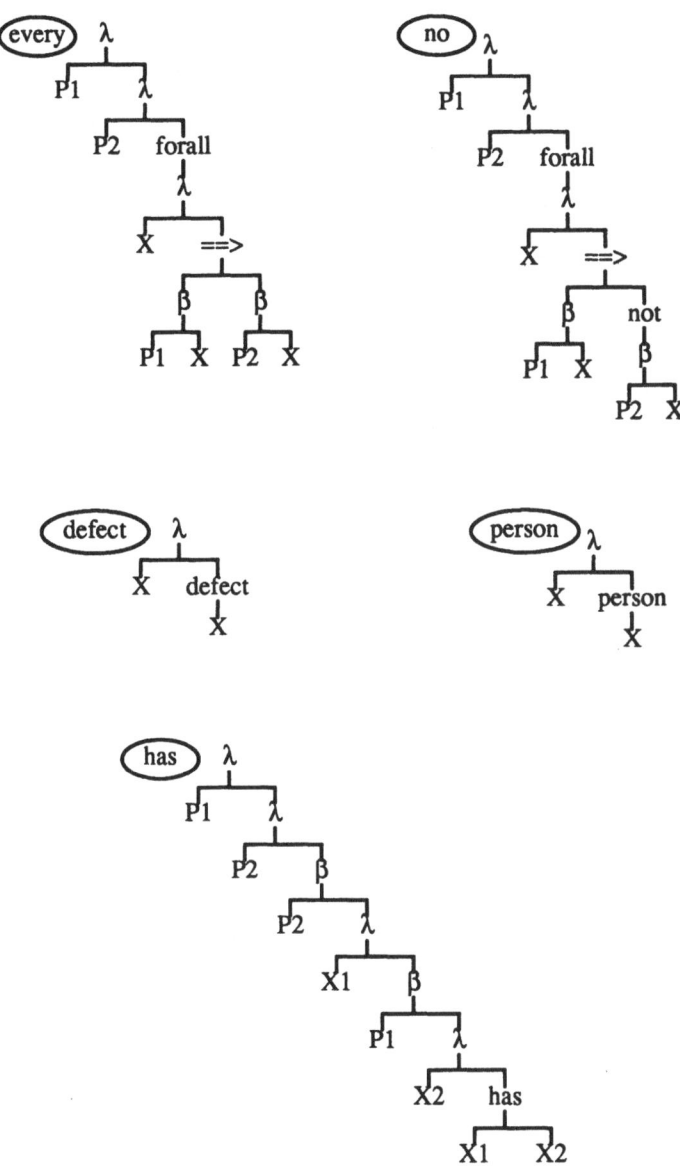

The grammar reads simply as follows:

sentence(S) --> noun-phrase(Q2), verb(W), noun-phrase(Q1),
        {apply(W,Q1,V), apply(V,Q2,S)}.

noun-phrase(Q) --> determiner(D), name(N), {apply(D,N,Q)}.

```
determiner(lambda(P1,lambda(P2,
      forall(lambda(X,implies(S1,not(S2)))))))
   --> [no], {apply(P1,X,S1), apply(P2,X,S2)}.
determiner(lambda(P1,lambda(P2,
      forall(lambda(X,implies(S1,S2)))))))
   --> [every], {apply(P1,X,S1), apply(P2,X,S2)}.

verb(lambda(P1,lambda(P2,S2)))
   --> [has], {apply(P2,lambda(X1,S1),S2),
      apply(P1,lambda(X2,has(X1,X2)),S1)}.

name(lambda(X,person(X))) --> [person].
name(lambda(X,defect(X))) --> [defect].
```

This grammar can be used either to analyze a given sentence into its corresponding formula, or to generate a sentence from a given formula. Here is, for instance, the formula corresponding to the sentence: "no person has every defect":

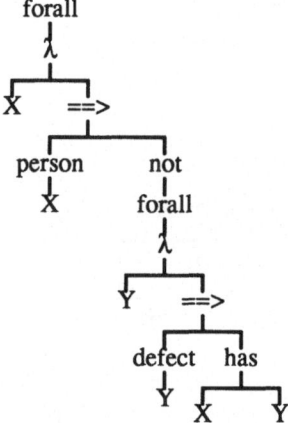

The SOLAR16 system uses a lambda-calculus representation for consulting an expert system in French. The parser has two phases: an analyser written by A. Colmerauer and an interface written by V. Dahl, with the formulas depicted in chapter 4, section 2 as final output.

The only other work known to us on Prolog implementations of grammars for lambda-calculus representations is by David Scott Warren (1983). He points out as an advantage of lambda-calculus representation its ability to directly represent meanings of phrases of all syntactic categories, while in other formalisms, such as the one described in chapter 4, section 2.3, phrases corresponding to some syntactic categories are represented by ill-formed formulas, whose use during parsing, of course, ultimately yields a final well-formed formula.

He also compares alternative implementations and concludes that, although the lambda-calculus is too expressive to be computationally tractable, part of it might be smoothly incorporated in the logic programming system itself, to allow the user to describe semantics in terms of a typed–lambda calculus while resorting to ordinary Prolog for syntax.

David H.D.Warren (1981a) on the other hand, gives arguments for representing lambda-expressions in an unextended Prolog. His discussion deals with logic programming as such rather than with natural language processing needs.

Chapter 5

Developing a Logic Grammar for a Formal Application

We have seen in the earlier sections how logic grammars may be developed for natural language applications. Now we shall consider the somewhat simpler problem of developing logic grammars for formal or programming languages. Many of the techniques used in the natural language applications are of course applicable. We begin with a grammar for logic programs; in a later chapter we consider a compiler for a simple traditional style programming language.

1. An Analyzer for Logic Programs

A logic program consists of a finite set of rules of the following form:

$$A :\text{-} B_1, \ldots, B_n. \qquad n \geq 0$$

When $n = 0$ a logic program rule may be written as:

$$A.$$

In a logic program rule the A and the B_i are called atoms and have the following form:

$$p(T_1, \ldots, T_n) \qquad n \geq 0$$

where p is a predicate symbol of arity n, $n \geq 0$, and the T_i are called terms. A term may be a variable, a constant, or constructed as follows:

$$f(T_1, \ldots, T_n) \qquad n \geq 0$$

where f is a function symbol and the T_i are terms. Variables are indicated by a string of letters and digits beginning with an upper case letter; constants, predicate symbols, and function symbols are indicated by a string of letters and digits beginning with a lowercase letter.

These definitions may be almost literally transcribed into a logic grammar which can be used to recognize logic programs. Let us begin with the definition of the nonterminal *logic_program* :

    logic_program --> [].
    logic_program --> rule, logic_program.

This specifies that a *logic_program* is either empty (designated by the empty list []) or consists of a *rule* followed by the rest of the *logic_program*. This corresponds to the definition of a logic program as a finite set of rules. We then go on to define the structure of a rule

rule --> head, tail, ".".

tail --> ":-", body.
tail --> [].

Here we deal with the two possibile forms for a logic program rule, the one with a nonempty body ($n > 0$) and the other with an empty body ($n = 0$). The nonterminal *tail* either consists of a :– symbol followed by a *body*, or it is empty. Symbols such as :– and the concluding period are represented as strings of symbols enclosed in quotation marks.

Continuing, we specify the nonterminals *head* and *body:*

head --> atom.

body --> atom, more_body.

A *head* consists of an *atom,* and a *body* consists of an *atom* possibly followed by *more_body:*

more_body --> ",", body.
more_body --> [].

The nonterminal *more_body* if not empty consists of a comma followed by another *body*.

Continuing in this top-down fashion we define the nonterminal *atom:*

atom --> predicate_symbol, atom_tail.

atom_tail --> [].
atom_tail --> term_tail.

An *atom* is a *predicate_symbol* followed by an *atom_tail* which is empty when the predicate symbol has arity $n = 0$, and otherwise is a *term_tail* defined by:

term_tail --> "(", term_list, ")".

term_list --> term, more_terms.

term --> variable.
term --> constant.
term --> function_symbol, "(", term_list, ")".

more_terms --> ",", term_list.
more_terms --> [].

The nonempty *term_tail* consists of a left parenthesis, followed by a *term_list* and a concluding right parenthesis. The definitions of *term* and *more_terms* should now be easy to follow.

Let us now define the remaining nonterminals which embody the specifications of predicate symbols, function symbols, variables, and constants. In order to do this we shall have to scan strings of letters and digits to make sure that they form the appropriate logic program symbols. This is a typical kind of lexical analysis necessary at the very lowest level of language processing, and also usually involves some extra-logical knowledge or extra-logical primitives.

```
predicate_symbol --> identifier.
function_symbol --> identifier.
constant --> identifier.
identifier --> small, lords.
variable --> large, lords
```

We call anything that begins with a lowercase (*small*) letter an *identifier*. A *variable* begins with an uppercase (*large*) letter. The remaining portion of an *identifier* and of a *variable* consists of a string (possibly empty) of letters or digits (*lords*).

```
lords --> [].
lords --> lord, lords.

lord --> letter.
lord --> digit.

letter --> [X], { letter_range(X) }.
digit --> [X], { digit_range(X) }.
large --> [X], { large_range(X) }.
small --> [X], { small_range(X) }.

letter_range(X) :- small_range(X).
letter_range(X) :- large_range(X).
digit_range(X) :- X < 58, X > 47 .
large_range(X) :- X > 64, X < 91.
small_range(X) :- X < 123, X > 96 .
```

The definitions of *letter, digit, large,* and *small* have right hand sides of the form:

```
[X], { condition(X) }
```

This defines one of these nonterminals as a character X which satisfies *condition(X)*. The conditions here are that the character falls into an appropriate range of characters. If one is using Prolog to implement this logic grammar, then the range is determined by an implementation defined mapping from external characters to internal integers. (Above, the definition of *letter_range*, *digit_range*, *small_range*, and *large_range* are dependent on the ASCII internal representation. Otherwise, one could define *digit* by the finite set of rules:

　　digit --> "0".
　　　　...
　　digit --> "9".

and *letter*, *small*, and *large* in a similar fashion.

We now have a logic grammar which can recognize logic programs by means of a query such as:

　　?- logic_program("mad(X):-logican(X).logician(lewis).",[]).

Exercises

1. Write logic grammar rules to specify a query which may be taken to be the symbol ?- followed by $B_1, \ldots, B_n$ where $n \geq 1$ and the B_i are atoms. □

2. Simple Lexical Analysis

The grammar for logic programs, although correctly specifying the syntax of a logic program, does not permit the logic programmer any flexibility in writing easily readable logic programs. Spaces are not permitted between symbols, nor are line breaks. The grammar is forced to mix "low level" concerns about how characters are grouped together with "high level" concerns about the structure of terms, atoms and rules. This can be remedied by introducing a separate analysis phase (also specified by logic grammar rules) which groups together characters into symbols relevant to the high level grammar. In our case, we wish to group characters into identifiers, variables, and the punctuation symbols *( ) , . :-* , and to allow spaces introduced for readability between these symbols. A query to analyze a logic program will then be the following conjunction of goals:

　　?- lexemes(Lex,Input,[]),logic_program(Lex,[]).

The input string *Input* is lexically analyzed, producing *Lex* and this is then passed to *logic_program* for checking to see that it is a properly formed logic program.

The definition of *lexemes* is simple: it consists of spaces followed by a *lexeme_list*. In this rule we make use of the possibility of attaching terms to nonterminals: the list of lexemes recognized by *lexeme_list* instantiates the logical variable *L* which is then passed to *lexemes:*

　　lexemes(L) --> spaces, lexeme_list(L).

We define spaces as follows:

```
spaces --> space, spaces.
spaces --> [].

space --> " ".
```

Additional rules may be added to the rule for *space* to deal with other ignorable symbols such as comments or end of line characters.

A *lexeme_list* is defined by:

```
lexeme_list([L|Ls]) --> lexeme(L), !, spaces, lexeme_list(Ls).
lexeme_list([]) --> [].
```

A *lexeme_list* is either empty, producing [] as its representation, or it consists of a *lexeme* which instantiates *L,* more spaces, followed by another *lexeme_list* instantiating *Ls*. The major *lexeme_list* consists of the list [*L* |*Ls*]. The symbol ! following *lexeme(L)* indicates that the lexeme *L* is unambiguously defined.

lexemes are defined by the following rules.

```
lexeme(I) --> id(I).
lexeme(V) --> var(V).
lexeme('(') --> "(".
lexeme(')') --> ")".
lexeme(',') --> ",".
lexeme('.') --> ".".
lexeme((:-)) --> ":-".
```

The remaining low level rules must be modified to transmit the characters which are recognized up to a point where they can be grouped into the appropriate symbols. We present the modified rules here:

```
lord(L) --> letter(L).
lord(D) --> digit(D).

lords([L|Ls]) --> lord(L), lords(Ls).
lords([]) --> [].

large(X) --> [X], { large_range(X) }.
small(X) --> [X], { small_range(X) }.
letter(X) --> [X], { letter_range(X) }.
digit(X) --> [X], { digit_range(X) }.
```

There is no change to the rules defining the ranges of characters.

The following rules transform appropriate strings of characters into function symbols representing variables and identifiers by means of the extra-logical predicate *name*. Given a string of characters "logician", name(L,"logician")

instantiates *L* to *logician*. Thus, by means of the following logic grammar rules
we produce *var('Xyz')* from *Xyz* and *id(mad)* from *mad*.

    var(var(V)) --> large(L), lords(Ls), { name(V,[L|Ls]) }.
    id(id(I)) --> small(S), lords(Ls), { name(I,[S|Ls]) }.

The remaining rules defining a logic program must now be modified to take into
account that they are analyzing symbols produced by lexical analysis rather than
the characters themselves. Here is the complete modified grammar:

    logic_program --> [].
    logic_program --> rule, logic_program.

    rule --> head, tail, ['.'].

    tail --> [:-], body.
    tail --> [].

    head --> atom.

    body --> atom, more_body.

    more_body --> [','], body.
    more_body --> [].

    atom --> predicate_symbol, atom_tail.

    atom_tail --> [].
    atom_tail --> term_tail.

    term_tail --> ['('], term_list, [')'].

    term_list --> term, more_terms.

    term --> variable.
    term --> constant.
    term --> function_symbol, ['('], term_list, [')'].

    more_terms --> [','], term_list.
    more_terms --> [].

    predicate_symbol --> identifier.
    function_symbol --> identifier.

constant --> identifier.

identifier --> [id(_)].
variable --> [var(_)].

A possible query for our grammar now is:

?- lexemes(L," mad(X) :- logican(X). logician(lewis). ",[]),
    logic_program(L,[]).

which would succeed with the list of lexemes:

L = [id(mad),'(',var(X),')',:-,id(logician),'(',var(X),')','.',
       id(logician),'(',id(lewis),')','.']

3. Structural Representation of a Logic Program

So far we have developed a purely syntactic grammar for logic programs which allows us to recognize whether a string of characters is or is not a logic program. We would now like to extend the grammar so that we get a representation of the structure of the logic program for any suitable string. There are many different ways of representing structure, so we must first decide how we want to represent the logic program, and then we will modify the grammar rules by adding arguments to nonterminal symbols so as to provide the chosen structure.

Since a logic program is a finite set of rules, it makes sense to represent a logic program as a set of rules. A set may be represented as a list, and we shall do so. Here now are the modified rules for *logic_program:*

logic_program([]) --> [].
logic_program([R|L]) --> rule(R), logic_program(L).

The empty *logic_program* is represented by [] while a nonempty *logic_program* is represented by the list [R |L] where R is the first rule and L is the set of remaining rules in the program.

A rule consists of a head and a tail. We shall represent this using the binary function symbol *rule* whose first argument is the head, and whose second argument is the tail of the rule.

rule(rule(H,T)) --> head(H), tail(T), ['.'].

The head H is determined by the definition of the nonterminal *head* and *tail* determines T.

Since a *head* is an *atom* we can let the structure of the latter determine the structure of the former. A nonempty *tail* inherits the structure of the body (we can dispense with the syntactic symbol *:-),* while a nonempty *tail* is represented by the symbol *true:*

head(H) --> atom(H).

tail(T) --> [(:-)], body(T).
tail(true) --> [].

The structure of a *body* should be a conjunction of atoms. We can modify the
rules for *body* and *more_body* as follows. The conjunction of atoms *B* is determined by the first atom *A*, obtained from *atom*. This *A* is then passed as the first
argument to *more_body* which conjoins any additional atoms with *A* to form *B*. If
there are no additional atoms (*more_body* is empty) then the first and second arguments of *more_body* are unified; otherwise, additional body is recursively found
in *B* and conjoined with *A* to form (*A,B*).

body(B) --> atom(A), more_body(A,B).

more_body(A,(A,B)) --> [','], body(B).
more_body(B,B) --> [].

An atom according to our definition begins with a predicate symbol of arity $n \geq 0$.
If $n > 0$, then there is a list of n terms (in parentheses) following the predicate symbol. We shall therefore use a ternary function symbol *atom* for our representation,
where the first argument will be the predicate symbol, the second its arity, and the
third the list of terms. We shall mark the arguments with the function symbols *p,
arity,* and *term_list* respectively. We determine *N* the length of the *term_list*
with the predicate *length*. The modified rules for *atom_tail* and *term_tail* should
be obvious.

atom(atom(p(P),arity(N),term_list(A))) -->
    predicate_symbol(P), atom_tail(A),
    { length(A,N) }.

atom_tail([]) --> [].
atom_tail(A) --> term_tail(A).

term_tail(T) --> ['('], term_list(T), [')'].

We then must choose representations for our terms. Let us represent terms which
are variables and constants with the unary function symbols *variable* and *constant*, and the recursive term by a ternary function symbol *structure* whose first
argument is a function symbol, second argument is its arity, and last argument is
the list of terms. We handle this in a manner similar to the way we defined a structure for *atom:* The rules for *term_list* and *more_terms* are then straightforward.

term(variable(T)) --> variable(T).
term(constant(T)) --> constant(T).
term(structure(f(F),arity(N),term_list(TList))) -->

```
        function_symbol(F), ['('], term_list(TList), [')'],
        { length(TList,N) }.

term_list([T|M]) --> term(T), more_terms(M).

more_terms(M) --> [','], term_list(M).
more_terms([]) --> [].
```

It only remains to define the modified rules for obtaining the representations of
predicate_symbol, function_symbol, constant, variable, and *identifier.* The following rules simply lift the symbols from the output of lexical analysis:

```
predicate_symbol(I) --> identifier(I).
function_symbol(I) --> identifier(I).
constant(I) --> identifier(I).
identifier(I) --> [id(I)].
variable(V) --> [var(V)].
```

Exercises

1. Define a structural representation for a query.
2. Define a predicate which will neatly display the structure of a *logic_program.*
   □

4. "Compiling" Logic Programs

Let us now give a different representation to our logic programs, one which will
form Prolog rules which may be asserted and then queried. In effect, we are compiling our logic grammar rules into a Prolog program which can then be executed.
Our logic grammar rules constitute what might be called "pure Prolog," with no
cuts, arithmetic expressions, extra-logical features, etc. Many of the rules are
identical to the representation which we obtained in the previous section; we shall
not bother commenting on those again.

A *logic_program* will now be a list of rules of the form *H:-T* where *H* is the head
and *T* is the tail of a rule. We introduce the symbol *true* as the tail of a unit rule.

```
logic_program([]) --> [].
logic_program([R|L]) --> rule(R), logic_program(L).

rule((H:-T)) --> head(H), tail(T), ['.'].

tail(T) --> [(:-)], body(T).
tail(true) --> [].

head(H) --> atom(H).

body(B) --> atom(A), more_body(A,B).
```

```
more_body(A,(A,B)) --> [','], body(B).
more_body(B,B) --> [].
```

The representation of an *atom* must now be generated in a form suitable for a Prolog program. We must take a list of symbols *[logician,X]* and convert them into the Prolog atom *logician(X)* for example. This will be accomplished by the predicate *form_structure* defined below.

```
atom(Atom) --> predicate_symbol(P), atom_tail(A),
        { form_structure(P,A,Atom) }.

atom_tail([]) --> [].
atom_tail(A) --> term_tail(A).

term_tail(T) --> ['('], term_list(T), [')'].
```

The representation of a *term* suitable for Prolog is given by the following rules. A structured term is handled similar to an *atom*.

```
term(T) --> variable(T).
term(T) --> constant(T).
term(T) --> function_symbol(F), ['('], term_list(TList), [')'],
        { form_structure(F,TList,T) }.

predicate_symbol(I) --> identifier(I).
function_symbol(I) --> identifier(I).
constant(I) --> identifier(I).
identifier(I) --> [id(I)].
variable(V) --> [var(V)].
```

The predicate *form_structure* is given by:

```
form_structure(P,A,Structure) :-
        append([P],A,L),
        Structure =.. L .
```

The symbol *P* is a predicate or function symbol which is formed into a single element list, appended to the list of terms *A* and then converted by =.. into a suitable Prolog structure.

5. Compiling Proof Tree Generation into Rules

In this section we shall be concerned as in the last with compilation of logic programs into Prolog programs, but with the difference that the Prolog clauses should now be compiled in such a way that proof trees of the solution to a query are constructed so that they can later be displayed and examined. For example, given the logic program:

```
mad(X) :- logican(X),loves(X,alice).
logician(lewis).
loves(lewis,alice).
```

and the query:

```
?- query(mad(Who)).
```

we obtain not only the solution *Who=lewis* but also a term representing the proof that *lewis* is mad which could be displayed in the form:

```
0: mad(lewis)
1: logician(lewis)
 2: true
1: loves(lewis,alice)
 2: true
```

We shall organize our new compiler so that the above logic program gets translated into the following Prolog clauses:

```
mad(X,proof(mad(X),[Proof_logician,Proof_loves])):-
    logician(X,Proof_logician),
     loves(X,alice,Proof_loves).
logician(lewis,proof(logician(lewis),[true])):-true.
loves(lewis,alice,proof(loves(lewis,alice),[true])):-true.
```

Note that in these Prolog clauses each atom has an extra argument which will be instantiated during evaluation to a proof of that atom. In general, a rule of the form

$$A :- B_1, \ldots, B_n. \quad n \geq 0$$

is translated into a Prolog clause of the form:

$$A' :- B'_1, \ldots, B'_n. \quad n \geq 0$$

where if

$$B_i = p_i(T_{i1}, \ldots, T_{im})$$

then

$$B'_i = p_i(T_{i1}, \ldots, T_{im}, ProofB_i).$$

If the head atom is of the form:

$$A = p_A(T_1, \ldots, T_k)$$

then

$$A'=p_A(T_1,...,T_k,proof(A,ProofList))$$

where

$$ProofList=[ProofB_1,...,ProofB_n]$$

A unit rule such as:

A.

where

$$A=p_A(T_1,...,T_k)$$

is translated into:

$$p_A(T_1,...,T_k,proof(p_A(T_1,...,T_k),[true])):-true.$$

The predicate *query* may be defined as follows:

```
query(X) :-
    form_proof_query(X,Proof,PX),
    call(PX),
    pretty(Proof).
```

Exercises

1. Define *form_proof_query*. □

The method we use in this compilation is similar to the one we used in the previous section, complicated however by the fact that we must now distinguish between constructed terms, atoms in the body, and the atom in the head. Earlier we made use of one predicate *form_structure* to form both terms and atoms no matter where they occurred. Now, formation of an atom in the body requires that we add a logical variable representing the proof of that atom to its list of terms, and formation of the atom in the head requires that we add the term *proof* (A ,*ProofList*) representing the use of the rule $A :-B_1,...,B_n.$ in the proof of the query to its list of terms. We use the predicate *form_structure* as before to form terms: it forms a term from a function symbol F and a list of terms A.

```
form_structure(F,A,Structure) :-
    append([F],A,L),
    Structure = L.
```

The predicate *form_atom* forms an *Atom* from a list consisting of a predicate symbol *P*, a list of terms *A* and a logical variable *AProof* which represents the proof of the atom. *form_atom* makes use of *form_structure*.

```
form_atom([P,A,AProof],Atom) :-
    append(A,[AProof],New),
    form_structure(P,New,Atom).
```

The predicate *form_head_atom* forms the head atom *H* from the predicate symbol *P*, the list of terms *A* and the list *TProof* of logical variables representing the proofs of the atoms in the *body*. The first argument in the term *proof (Struct ,TProof)* which instantiates *HProof* is constructed as a result of *form_structure (P ,A ,Struct)*. The atom *H* is formed using *form_atom*.

```
form_head_atom([P,A],TProof,H) :-
    HProof = proof(Struct,TProof),
    form_structure(P,A,Struct),
    form_atom([P,A,HProof],H).
```

Here follows the logic grammar for the new compiler. The nonterminal *body* also has a second argument representing the list of proofs.

```
logic_program([]) ::= [].
logic_program([R|L]) ::= rule(R), logic_program(L).

rule((H :- T)) ::= head(Head),  tail(T,TProof), ['.'],
        { form_head_atom(Head,TProof,H) }.

tail(T,TProof) ::= [(:-)], body(T,TProof).
tail(true,[true]) ::= [].
```

Note that the nonterminal *tail* now has two arguments, the first representing the tail of body atoms or *true,* the second representing the list of logical variables representing the proofs of the atoms in the tail or the trivial proof [*true*].

```
head(Head) ::= atom(Head).

body(B,[AProof|MProof]) ::=
        atom([P,Terms]), more_body(BodyAtom,B,MProof),
        { form_atom([P,Terms,AProof],BodyAtom) }.

more_body(A,(A,B),BProof) ::= [','], body(B,BProof).
more_body(B,B,[]) ::= [].
```

In the rule for *body* the first atom consisting of the predicate symbol *P* and the list of terms *Terms* is recognized. *BodyAtom* is formed from these and the logical variable *AProof*. Any further atoms in the body are recognized by *more_body* which given *BodyAtom* returns the complete body *B* and *MProof*, the list of proof variables of the rest of the body. The list of proofs is then constructed as [*AProof* |*MProof*]. The rules for *more_body* have a third argument which represents the proof list of the remaining body.

    atom([P,A]) ::= predicate_symbol(P), atom_tail(A).

The argument of *atom* now is the pair consisting of *P*, the predicate symbol, and *A*, the list of terms. The remaining rules are as before.

    atom_tail([]) ::= [].
    atom_tail(A) ::= term_tail(A).

    term_tail(T) ::= ['('], term_list(T), [')'].

    term(T) ::= variable(T).
    term(T) ::= constant(T).
    term(T) ::= function_symbol(F), ['('], term_list(TList), [')'],
        { form_structure(F,TList,T) }.

    term_list([T|M]) ::= term(T), more_terms(M).

    more_terms(M) ::= [','], term_list(M).
    more_terms([]) ::= [].

    predicate_symbol(I) ::= identifier(I).
    function_symbol(I) ::= identifier(I).
    constant(I) ::= identifier(I).

Exercises

1. Define a logic grammar for a simple logic grammar notation where the only thing that can appear within the braces { } are calls of pure logic predicates as defined above. First write a grammar simply to recognize correct logic grammars, then extend this to a compiler from logic grammar rules to Prolog clauses.

2. Define a logic grammar for *all* of a version of Prolog. (This may be difficult and may require some knowledge of parsing techniques for programming languages.) □

Bibliographic Commentary for Part II

The step-by-step analyzer development presented in chapter 3 evolved from work described in Dahl (1981). The logical form internal representations for querying databases presented in chapter 4 were first covered in Dahl (1982, 1983). Further work on quantification can be found in Pique (1982) and Saint-Dizier (1985, 1986). Another early natural language interface system is described in Wallace (1984).

M.McCord (1982, 1981) generalized the traditional notions of determiners and modifiers so that every syntactic item has a determiner, and all types of phrases are reduced to a simple structure called a *syntactic item*, of the form:

    syn (Features, Determiner, Predication, Modifiers)

where the *Determiner* determines how a syntactic item can modify another one. *Features* contains syntactic and morphological information about the item; *Predication* is the central predication of the *item* (e.g. involving the head noun in a common noun phrase); and *Modifiers* is the list of all modifiers of the given item, listed in their surface order. Each modifier is again a *syn*.

Thus, all the modifiers are laid out in the syntactic structure, to look at before trying to decide what the correct order is. After a sentence is analyzed into a *syn* tree, this tree is reshaped taking into account the relative scopes of its nodes (priority levels for determiners are used here), and finally, a third phase actually applies the determiners to get the final formula. Dahl and McCord (1983) further generalize this treatment to deal also with scoping problems involving coordination.

Another recent approach organizes semantic interpretation into a single stage in which processes such as sense selection, slot filling and other types of modification are interleaved McCord (1984). A tutorial survey of the uses of logic in database systems can be found in Gallaire et al. (1984).

Since the development of SOLAR16, primitives implementing safe negation as failure have been incorporated directly into Prolog versions. Notably, the one used in IC-Prolog (Clark et al. 1980) reports a control error for each attempt to execute an unsafe negated expression. It seems, however, more intelligent to postpone its execution in the hope of its becoming safe, and only report an error when this has proved impossible. This is the course taken in SOLAR16, as well as in the Prolog versions (Colmerauer et al. 1983; Giannesini et al. 1986), and in Mu-Prolog (Naish 1983).

CHAT80 (Warren 1981) also treats negation in this way, although the negated formulas allowed are more restricted. Futher details on Prolog negation are referenced at the end of this chapter.

The "compilers" for pure Prolog and proof tree generation which we have described in chapter 5 are compiled analogues of some of the metainterpreters described in chapter 19 of Sterling and Shapiro (1986). Good sources for general information on compilation techniques (including lexical analysis) are Aho and

Ullman (1972, 1977), Aho et al. (1986). See also our chapter on definite clause translation grammars (chapter 9) for more information about compilation using logic grammars and further references on the topic.

PART III

Different Types of Logic Grammars - What Each Is Useful For

Chapter 6

Basic Types of Logic Grammars

Having discussed how to translate natural language into logical form and logic programs into Prolog, let us examine some of the existing types of logic grammars. While doing so we will gain further insight as to how to go about writing one. More sophisticated types of logic grammars are reviewed in chapters 7, 8, and 9. From a theoretical point of view, the power of the types of grammars we shall introduce is similar in that they can all serve to describe type-0 languages. From a practical point of view, however, their expressive possibilities differ.

This section illustrates this point by studying how easily and concisely each of these formalisms allows us to describe those rules involving constituent movement and elision. We have very briefly presented some ideas on movement for producing interrogative clauses (cf. chapter 3, section 6). We now develop the discussion around the following example for relativization.

Let us consider the noun phrase:

>   the man that John saw

This noun phrase can be thought of as the surface expression of the more explicit form:

>   the man [John saw the man],

where the second occurrence of "the man" has been shifted to the left and subsumed into the relative pronoun "that".

Let us consider a simpler grammar than the one developed earlier, for (very restricted) sentences described by the following grammar:

>   (1) sentence --> noun_phrase,verb_phrase.
>
>   (2) noun_phrase --> determiner,noun,relative.
>   (3) noun_phrase --> proper_name.
>
>   (4) verb_phrase --> verb.
>   (5) verb_phrase --> trans_verb,direct_object.
>
>   (6) relative --> [].
>
>   (7) direct_object --> noun_phrase.
>
>   (8) determiner --> [the].

(9) noun --> [man].

(10) proper_name --> [john].

(11) verb --> [laughed].

(12) trans_verb --> [saw].

We shall successively modify this grammar (referred to as G in all that follows) in order to describe the relativization process in three different logic grammar formalisms.

1. Metamorphosis Grammars

Metamorphosis grammars (MGs), the first type of logic grammars, were developed by Alain Colmerauer (1975). MG rules have the form:

Sα-->β

where S is a nonterminal (logic) grammar symbol, α is a string of terminals and nonterminals[1] and β is a string of terminals, nonterminals and procedure calls. This is the type of logic grammar we have already been using in sections 5 and 6 of chapter 3.

Within MGs, all we need to do to modify grammar G so that it will treat the noun_phrase "the man that john saw" appropriately, is to add the rules:

(5') verb_phrase --> moved_dobj,transitive_verb.

(6') relative --> relative_marker,sentence.
(13) relative_marker,noun_phrase,moved_dobj -->
                        rel_pronoun,noun_phrase.
(14) relative_pronoun --> [that].

The modified version of *G* will be called *G'*. Figure 1 depicts the derivation graph for our sample noun phrase "the man that John saw". We abbreviate some of the grammar symbols. Rule numbers appear as left-hand side labels.

Of course, for such a parse to be of any use, we need to construct a representation for the sentence while we parse it. But for the time being we shall ignore symbol arguments in order to concentrate upon the particular problem of moving constituents.

---

[1] The actual Prolog implementation of MGs requires in fact that α be a string of terminals alone, but we shall disregard this restriction since it has been shown (Colmerauer 1975) to involve no substantial loss with respect to the full MG form.

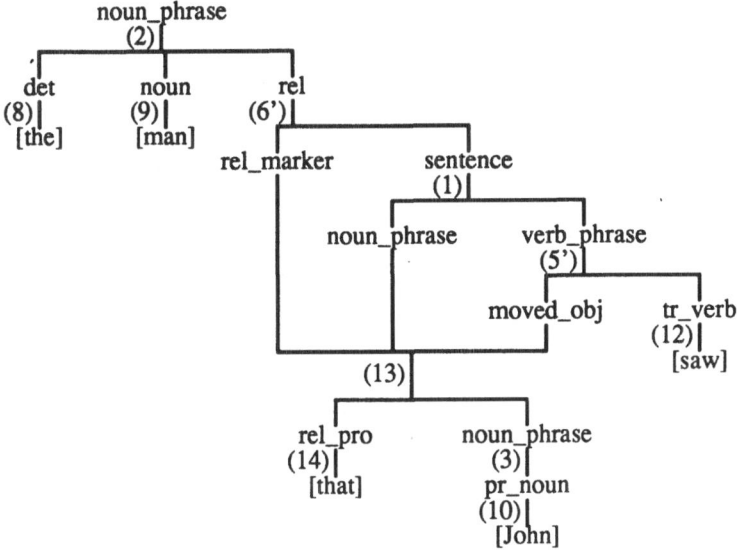

Figure 1.
   MG derivation graph for "The man that John saw",
   corresponding to grammar G'.

2. Definite Clause Grammars

Definite clause grammars (Pereira and Warren 1980), included in DEC-10 Prolog (Pereira et al. 1978), are a special case of MGs, where rules are restricted to a single, nonterminal symbol on the left-hand side, i.e.:

$$S \dashrightarrow \beta$$

The main motivation for introducing DCGs was ease of implementation coupled with no substantial loss in power, in the sense that because symbols may include arguments, DCGs can also describe type-0 languages – although less straightforwardly.

In terms of DCG rules, the simplest possible modification to our original grammar G is to allow a direct object to be elided, e.g., by adding the rule:

   (7') direct_object --> [].

But, because this rule lacks the contextual information found in (13), a direct object is now susceptible to being elided even outside a relative clause. In order to prevent it, one technique is to control rule application by adding extra arguments. In our example, we only need to add a single argument that we carry within the *sentence*, *verb_phrase* and *direct_object* symbols, and that takes the value "[]" if the direct object in the verb phrase of the sentence is not elided, and the value "elided" if it is. The modified rules follow:

(0) sentence --> sent([]).
(1) sent(E) --> noun_phrase,verb_phrase(E).

(4) verb_phrase([]) --> verb.
(5) verb_phrase(E) --> transitive_verb, direct_object(E).

(6') relative --> relative_pronoun,sent(E).

(7) direct_object([]) --> noun_phrase.
(7') direct_object(elided) --> [].

(13) relative_pronoun --> [that].

Figure 2 shows the DCG derivation tree for "The man that John saw laughed". Substitutions of terms for variables are shown as right-hand side labels.

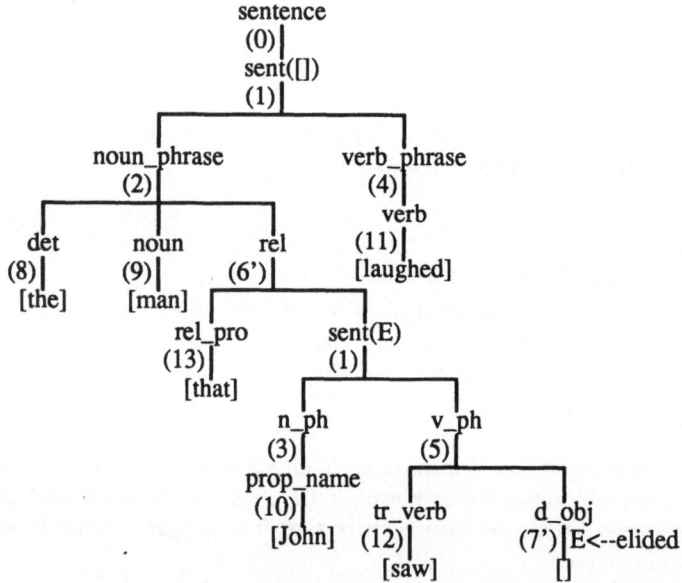

Figure 2.
    DCG derivation tree for the sentence
    "The man that John saw laughed"

3. Extraposition Grammars

Extraposition grammars (XGs) introduced by F. Pereira (1981) allow us to refer to unspecified strings of symbols in a rule, thus making it easier to describe left extraposition of constituents.

XGs allow rules of the form

$$s_1 \cdots s_2 \; etc. \; s_{k-1} \cdots s_k \rightarrow r$$

where the " $\cdots$ " specify skips (i.e., arbitrary strings of grammar symbols), and r and the s_i are strings of terminals and nonterminals.

The general meaning of such a rule is that any sequence of symbols of the form

$$s_1 x_1 s_2 x_2 \; etc. \; s_{k-1} x_{k-1} s_k$$

with arbitrary x_i's, can be rewritten into $r x_1 x_2 \cdots x_{k-1}$ (i.e., the s_i's are rewritten into r, and the intermediate skips (x_i's) are rewritten sequentially to the right of r. For treating relative clauses, for instance, the XG rule:

> relative_marker. . .complement --> [that].

allows us to skip any intermediate substring appearing after a relative marker in the search for an expected complement, and then to subsume both marker and complement into the relative pronoun ''that'', which is placed to the left of the skipped substring.

While, as we have seen, MGs express movement by actually moving constituents around, DCGs must carry all information relative to movements within extra arguments. XGs, on the other hand, can capture left extraposition in an economical fashion by actually skipping intermediate substrings rather than shifting the constituents that follow. Thus, our initial grammar can be modified to handle relativization simply by adding the XG rules:

> (6') relative --> relative_marker, sentence
> (13) relative_marker ... direct_object --> relative_pronoun.
> (14) relative_pronoun --> [that].

Figure 3 shows the XG derivation tree for ''The man that John saw laughed''.

Similarly, the grammars shown in chapter 3, section 6 can generate much simpler derivation graphs if we merely replace rule PR by the XG rule:

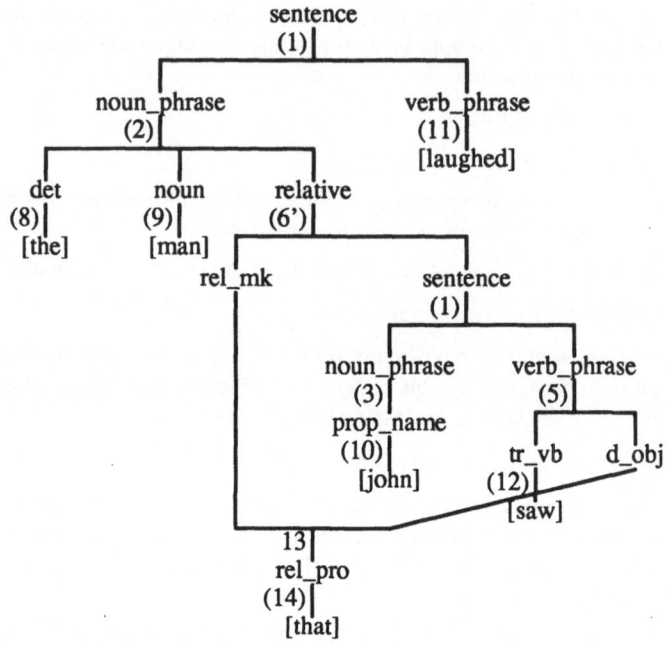

Figure 3.
    XG derivation tree for
    "The man that John saw laughed"

    PR') wh_1(K), . . . , modifier(K, S1, S2) --> [dónde].

The skeleton derivation would now be as shown in figure 4.

4. Discussion

The three formalisms covered in this chapter are historically the earliest, and, as we have said, equivalent in terms of computational power. Which one to choose for a particular application is largely a matter of personal taste. If the grammar author does not mind argument proliferation, DCGs might be adequate. If the author prefers to specify movement phenomena directly, MGs or XGs might be preferred. If a moved constituent has to skip over constituents using several MG rules successively, XGs might be able to achieve the same result with just one rule, and might therefore be preferred.

Exercises

1. Adapt the Spanish grammar developed in chapter 3 into English

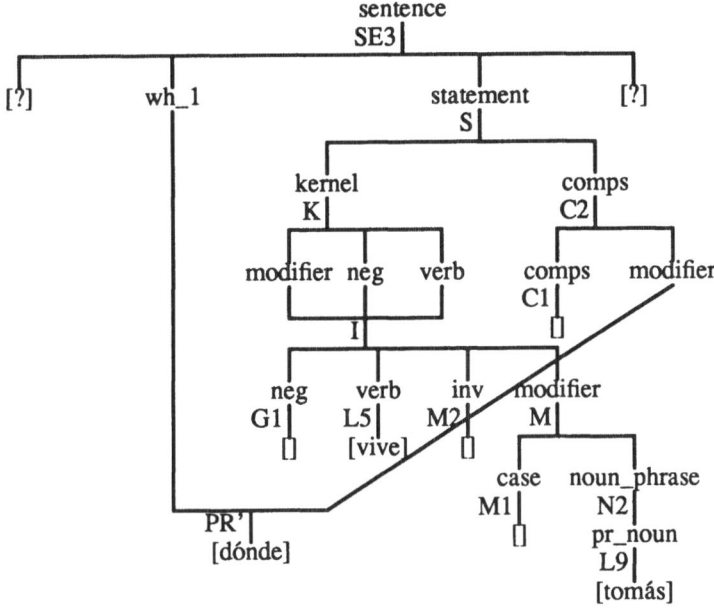

Figure 4.
XG skeleton derivation graph for
"Dónde vive Tomás?"

    a. as a metamorphosis grammar
    b. as a definite clause grammar
    c. as an extraposition grammar

Include negation, quantified noun phrases, and a treatment of both relative and
interrogative clauses, in as complete a manner as possible. ☐

Chapter 7

Building Structure

Several variants of logic grammars have been proposed lately, sometimes motivated by ways of solving particular natural or formal language processing problems, sometimes by the need of providing easier or more general ways of processing grammars.

One of the useful ideas born from these developments is that there is no need for the user to laboriously describe the construction of a parsing tree in his rules, or even to detail the construction of his meaning representations.

1. Parse Tree Construction

Building a parse tree is a boring but not difficult task, at least in the simplest grammar formalisms, like DCGs. Extra arguments in non-terminal symbols can be used, where structures are progressively built up during unification, as grammar rules are applied. The following grammar, taken from Pereira and Warren (1980), illustrates this:

```
sentence(s(NP,VP)) --> noun_phrase(NP), verb_phrase(VP).

noun_phrase(np(Det,Noun,Rel)) --> determiner(Det), noun(Noun),
    rel_clause (Rel).
noun_phrase(np(Name)) --> name(Name).

verb_phrase(vp(TV,NP)) --> trans_verb(TV), noun_phrase(NP).
verb_phrase(vp(IV)) --> intrans_verb(IV).

rel_clause(rel(that,VP)) --> [that], verb_phrase(VP).
rel_clause(rel(nil)) --> [].

determiner(det(W)) --> [W], {is_determiner(W)}.

noun(n(W)) --> [W], {is_noun(W)}.

name(name(W)) --> [W], {is_name(W)}.

trans_verb(tv(W)) --> [W], {is_trans(W)}.

intrans_verb(iv(W)) --> [W], {is_intrans(W)}.
```

Examples of Prolog rules from the associated dictionary are:

```
is_determiner(every).
is_noun(man).
```

is_name(mary).
is_trans(loves).
is_intrans(lives).

With this grammar, the query:

?-parse(sentence(Parse_tree)),[every,man,lives]).

results in the parse tree of figure 5.

Parse_tree =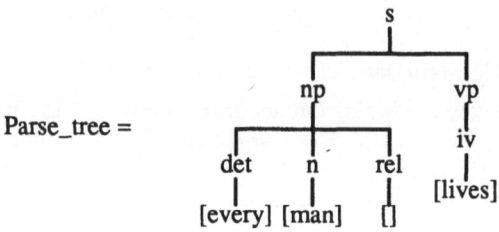

Figure 5.
    A simple parse tree

This buildup task is mechanical enough that it can be implemented invisibly to the user. A program can be written for adding the extra arguments for structure-building automatically.

2. Meaning Representation Buildup

Building a meaning representation, on the other hand, involves ingenuity to put together the appropriate pieces with respect to our particular target language. The following grammar, for instance, maps the subset of English described by the previous grammar into first-order logic:

    sentence(P) --> noun_phrase(X,P1,P), verb_phrase(X,P1).

    noun_phrase(X,P1,P) --> determiner(X,P2,P1,P), noun(X,P3),
        rel_clause(X,P3,P2).
    noun_phrase(X,P,P) --> name(X).

    verb_phrase(X,P) --> trans_verb(X,Y,P1),
             noun_phrase(Y,P1,P).
    verb_phrase(X,P) --> intrans_verb(X,P).

    rel_clause(X,P1,P1 & P2) --> [that],verb_phrase(X,P2).
    rel_clause(X,P,P) --> [].

    determiner(X,P1,P2,all(X):(P1 ==> P2)) --> [every].

determiner(X,P1,P2,exists(X):(P1 & P2)) --> [a].

noun(X,man(X)) --> [man].
noun(X,woman(X)) --> [woman].

name(john) --> [john].

trans_verb(X,Y,loves(X,Y)) --> [loves].

intrans_verb(X,lives(X)) --> [lives].

The query:

?-sentence(P,[every,man,lives],[]).

now unifies P to the tree or meaning representation of figure 6.

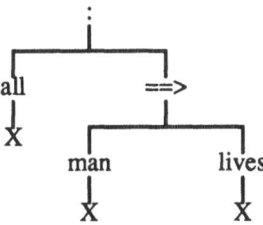

Figure 6.
   The meaning of "Every man lives"

We can instead obtain typed formulas with "set_of" as the unique quantifier, by replacing the last seven rules above by:

determiner([X,T],P1,P2,set_of(X,T,P1&not(P2),S)&
          card(S,0)) --> [every].
determiner([X,T],P1,P2,set_of(X,T,P1&P2,S)&
          card(S,C)&gt(C,0)) --> [a].

noun([X,human],woman(X)) --> [woman].
noun([X,human],man(X)) --> [man].

name([mary,human]) --> [mary].
name([tweety,bird]) --> [tweety].

trans-verb([X,human],[Y,_],loves(X,Y)) --> [loves].
intrans-verb([X,_],lives(X)) --> [lives].

The same query above now unifies P with the structure of figure 7.

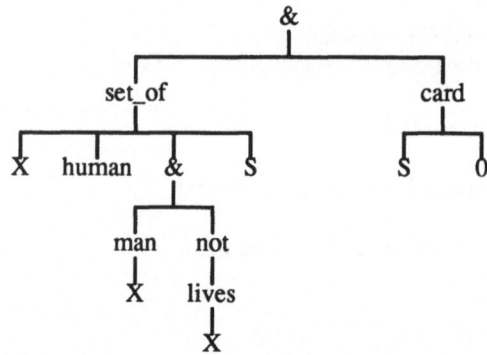

Figure 7.
    The set representation of "Every man lives"

Figures 8, 9, and 10 give the representations respectively obtained by the three above grammars, for the sentence "Every man that lives loves a woman".

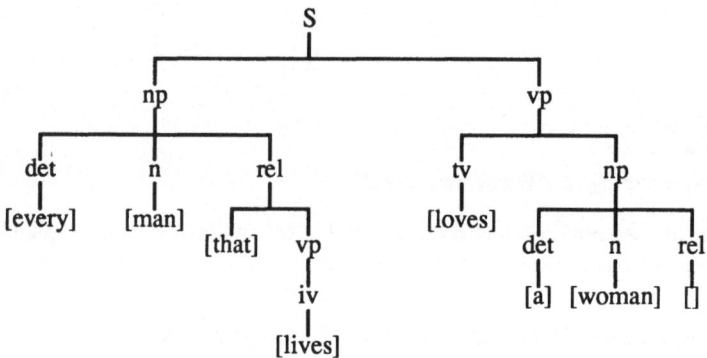

Figure 8.
    Parse tree representation of
    "Every man that lives loves a woman"

3. Automating Syntactic and Semantic Structure Buildup

Notice that we could easily combine the above grammars to obtain *both* a parse tree and a meaning representation of the sentence parsed. It suffices to have an extra argument of each of these, e.g.:

        sentence(P,s(NP,VP)) --> noun_phrase(X,P1,P,NP),
            verb_phrase(X,P1,VP)

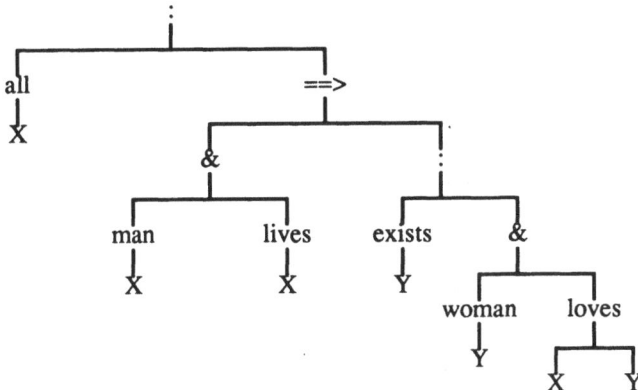

Figure 9.
    First order logic representation of
    "Every man that lives loves a woman"

where the last argument in each symbol deals with parse tree buildup and the other arguments, with meaning representation.

It is clear that, if we could somehow hide these extra arguments from the user, our grammar descriptions would become more legible. Hiding parse-tree arguments, as we have said, is quite straightforward.

Meaning representations in more complex grammars, however, is not a trivial matter, as these representations usually follow neither the history of rule application nor the surface ordering of constituents, and can therefore not be described in sequential terms.

Arriving at an appropriate meaning representation may involve complex transformations, change of modifier scopes, etc. (cf. our discussion on quantifier scoping in chapter 4, section 2.3). It may not be possible to completely automate this process, but it is possible to automate some of it, by giving the user a means for specifying guidelines which the system will consult in order to construct the final representation. We should allow for these guidelines to be set in a modular fashion, so that modifications to the representations obtained can be achieved with just local changes that are relatively independent from syntax.

Joint research with M.McCord on coordination (Dahl and McCord 1983) yielded as a by-product a new grammar formalism, *modifier structure grammars,* which automates the construction of both parse tree and (in the modular fashion described above) logical representation. It also separates the semantic components of rules from the syntactic one and automates quantifier scoping as well as coordination. The next chapter describes this approach.

Simultaneously and independently, restriction grammars were developed (Hirschman and Puder 1984), which also automate parse tree construction, although not

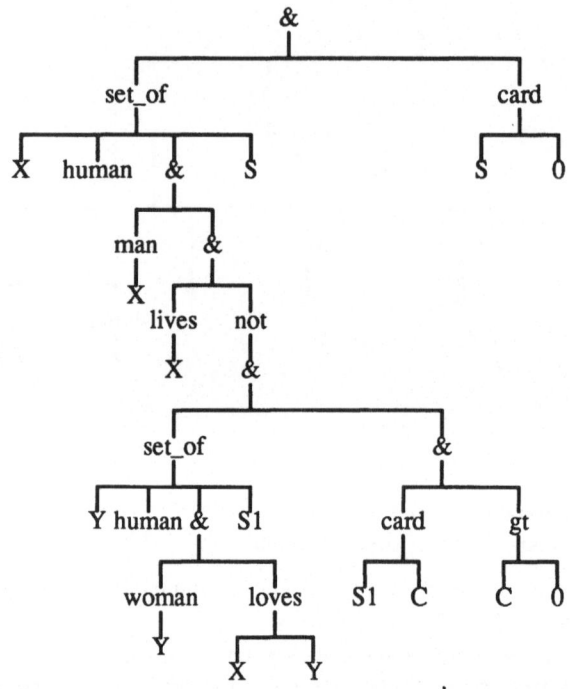

Figure 10.
   Set representation of "Every man that lives loves a woman"

logical representation, within an augmented context-free framework. Abramson later elaborated on the idea of automatic buildup of logical form in his "Definite Clause Translation Grammmars" (Abramson 1984) (cf. chapter 9).

A generalization of XGs, called *discontinuous grammars*, where skips can be arbitrarily repositioned, was developed by Dahl (1984b), and some of its implementation issues were examined jointly with McCord (1982), with Abramson (1984) and by Popowich (1985). Sabatier (1985) has developed *puzzle grammars*, in which languages are defined through trees instead of rewrite rules. These trees are "glued" together during parsing, as elements of a puzzle. The next three chapters describe some of the most recently developed formalisms.

Chapter 8

Modifier Structure Grammars

1. Separation of Syntax and Semantics

As we saw in the preceding chapter, it is easy to automate syntactic tree buildup. It is not as easy, but often more desirable, to automate the buildup of semantic structure. Both features were introduced in joint research with McCord (Dahl and McCord 1983), which was motivated by treating coordination (constructions with "and", "or", "but", etc.) in a metagrammatical fashion.

MSG rules clearly separate, in the rule format, the syntactic from the semantic elements that the rule contributes to the total syntactic and semantic structure. An MSG rule has the following form:

A:Sem --> B

where A --> B is an XG rule and Sem is a term called a *semantic item*, which plays a role in the semantic interpretation of a phrase analyzed by application of the rule. The semantic item is (as in McCord [1981]) of the form

Operator-LogicalForm

where, roughly, LogicalForm is the part of the logical form of the sentence contributed by the rule, and Operator determines the way in which this partial structure combines with others. Sem may be a "trivial" Sem if nothing is contributed.

When a sentence is analyzed, a structural representation, in tree form, called "modifier structure" is automatically formed by the parser. Each of its nodes contains not only syntactic information, but also the semantic information Sem supplied in the logical form of the sentence (this contribution is for the node alone, and does not refer to the daughters of the node, as in Gazdar's approach [Gazdar 1981]).

2. Quantifier Rescoping

The semantic interpretation component first reshapes this tree into another modifier structure tree where the scoping of quantifiers is closer to the intended semantic relations than to the (surface) syntactic ones. It then takes the reshaped tree and translates it into a logical form. The modifiers actually do their work of modification in this second stage, through their semantic items. It should be noted that the addition of simple semantic indicators within grammar rules contributes to the maintainance, from the user's point of view, of a simple correspondence between syntax and semantics. This is similar in intention to the rule-by-rule hypothesis introduced in (Bach 1976) (to each syntactic rule corresponds a semantic rule), but is differently realized: instead of a rule-to-rule correspondence, we have a correspondence between each nontrivial expansion of a nonterminal and a logical operator. That is, each time the parser expands a nonterminal symbol into a (nonempty) body, a logical operator labels the expansion and will be later used by the semantic component, interacting with other logical operators found in the parse tree obtained. The complexity of dealing with quantifier scoping and its interaction with coordination is screened away from the user.

3. Coordination

Coordination is one of the most difficult natural language processing problems, because it spans a wide range of grammatical constituents, or even nonconstituent fragmants, and ellipsis can occur in the items conjoined. A typical example is Wood's:

John drove the car through and completely demolished a window

from which a machine would have to reconstruct the ellided constituent "John" in the second conjunct, and "a window" in the first. Moreover, it would have to produce a meaning representation in which the window referred to in the first coincided with the one referred to in the second. In other words, it would not be enough to simply reconstitute the missing string (even assuming we could easily identify it) and analyze "a window" from its two occurrences. This would not only mean redundant work, but would result in two different variables (two different windows) being introduced by the two "a" quantifiers.

The interaction between quantifier scoping and coordination is also automated in MSGs, however. Basically, the user describes a specific grammar, in terms of MSG rules, which does not mention coordination at all. The interpreter has a general facility for treating certain words as ''demons'' (cf. Winograd [1972]), which trigger a backing up in the parse history that will help reconstruct ellisions and recognize the meaning of the coordinated sentence.

This proceeds in a manner similar to that of the SYSCONJ facility for augmented transition networks (Woods 1973; Bates 1978), except that, unlike SYSCONJ, it can also handle embedded coordination, coordination of more than two elements, and interactions with extraposition. The use of modifier structures and the associated semantic interpretation components, moreover, permits in general a good treatment of scoping problems involving coordination. As an example of the effects of this treatment, for the grammar tested in Dahl and McCord (1983), the sentence:

Each man and each woman ate an apple

is given the logical form

each(X,man(X),exists(Y,apple(Y),ate(X,Y)))
        & each(X,woman(X),exists(Y,apple(Y),ate(X,Y))),

whereas the sentence

A man and a woman sat at each table

is given the form

each(Y,table(Y),exists(X,man(X),sat_at(X,Y))
      & exists(X,woman(X),sat_at(X,Y))).

4. Implementation

The MSG formalism was implemented through a Prolog-10 (Pereira et al. 1978) interpreter that takes all the responsibility for the syntactic aspects of coordination (as with SYSCONJ) and a semantic interpretation component that produces logical forms from the output of the parser and deals with scoping problems for coordination. The complete system is, although interpreted rather than compiled, reasonably efficient (cf. timings of the sample grammar in Dahl and McCord [1983]). It is also possible to compile MSGs into XGs, which in turn can be compiled into Horn clauses. But although the part of making explicit the structure buildup is fairly direct, it does not seem easy to compile in a way that treats coordination automatically.

5. Discussion

The structure automatically produced from an MSG-parsed sentence can be considered an annotated phrase structure tree, but the underlying grammar is not necessarily context-free. The rules accepted are generalized type-0 rules that may include skips (in view, for instance, of left extraposition). Semantic interpretation is guided through the semantic items, local to each rule, which help resolve scoping problems. Because the semantic information about the structure being built up is described modularly in the grammar rules, it becomes easier to adapt the parser to alternate domains of application: modifying the logical representation obtained need only involve the semantic items in each rule.

It also seems important, in view of expressiveness and conciseness, that MSGs do not preclude context sensitive rules, transformations, or skips. Most approaches that advocate the restriction to context-free grammars end up supplementing them with features such as restrictions, local constraints, rule schemata, etc. to make up for their lack of expressiveness. Since MSGs can also treat in particular context-free rules, they will also be appropriate for those cases in which only these are required, and at the same time they will not need to resort to extra features if more than context-free power is desired.

In Dahl and McCord (1983) the observation was made that the MSG formalism, outside the specific problem of coordination that motivated its development, seems of interest in its own right, perhaps even outside natural language processing, because the notions of syntactic structure and semantic interpretation are largely implicit in the formalism itself, so that less burden is put on the programmer. Harvey Abramson acted upon this remark, and isolated these notions into the

Definite Clause Translation Grammar formalism, which retains the modular treatment of semantics and the automation of structure buildup while giving up context
sensitive, transformational, and discontinuous rules (rules with skips), as well as
the automation of quantifier-scoping and of coordination treatment. The next
chapter presents DCTGs and shows their usefulness for problems outside natural
language processing.

Exercises

1. Compare different types of coordinated sentences with ellisions and discuss
   what representations should be obtained, expecially taking quantifier scope
   into account.

2. Write a logic grammar that will produce adequate representations for just one of
   these types of sentences. □

Chapter 9

Definite Clause Translation Grammars and their Applications

1. Definite Clause Translation Grammars

DCGs, as we have seen, provide a general mechanism for grammatical computation. One could, for example, define a DCG for a subset of a natural language which would analyze sentences, form a derivation tree, check for agreement between subject and verb, translate input sentences into some logical form, and so on. The trouble with such an all-inclusive DCG, however, is that with an increasing number of arguments attached to the nonterminal symbols, the rules become difficult to read and/or modify. In order to change something in the logical form construction, for example, one would have to be very careful to adjust the right argument in each affected nonterminal symbol: one has to make semantic changes within a rule which incorporates in a nonmodular fashion both syntax and semantics. It is all too easy to become confused and make a significant change to the behavior of a grammatical specification. Furthermore, as we have seen in chapter 7, section 3, and in chapter 8, both the parse tree formation, and the formation of semantic structure, are mechanizable, and it is unnecessary to impose on the grammar writer any mechanical task.

Definite Clause Translation Grammars (or DCTGs) were devised by Abramson to overcome some of these problems with DCGs. Parse trees are formed automatically as a result of a successful derivation. Semantics are embodied in a set of Horn clause rules attached to each nonterminal node of the parse tree. These rules guide the computation of various semantic properties of the node in terms of computations of semantic properties of subtrees of the node. This attachment of a semantic rule or rules is consistent with what in natural language research has been termed the "rule-to-rule hypothesis" by Bach (1976), and described more recently in Gazdar et al. (1985) as follows: "The [hypothesis] maintains that each syntactic rule in a grammar for a natural language is associated with a semantic rule which determines the meaning of the constituent whose form the syntactic rule specifies."

DCTGs were, however, devised not only for the pragmatic reason of overcoming the difficulties of using DCGs in complex applications, but also as a logical version of what are known as attribute, or property grammars, originally introduced by Knuth. In these grammars, syntax is specified by context-free rules. Semantics are specified by attributes or properties attached to nonterminal nodes in derivation trees and by functional rules for computing the properties of nodes in terms of values "inherited" from above in the tree and from values "synthesized" from subtrees. The specification of an attribute grammar is declarative, and it is a computationally complex problem to decide whether the properties of the root node of a derivation tree may be computed from the given rules without getting into an infinite loop.

In the DCTG formalism we attach a set of Horn clauses to the nonterminal nodes of a derivation tree. We do not distinguish between inherited and synthesized attributes in the notation, as the logical variable makes this distinction largely unnecessary. (Such distinctions in the nature of mode declarations might be useful, however, in helping a Prolog compiler to generate efficient code.) The execution of the semantic Horn clauses is specified by an interpreter and uses SLDNF resolution.

Some of the examples we give are based on examples familiar to those who know the literature on attribute grammars. This is deliberate and meant to emphasize that DCTGs are a logical version of attribute grammars with SLDNF resolution as the execution control.

Example 1.

bit ::= "0".

bit ::= "1".

bitstring ::= [].

bitstring ::= bit, bitstring.

number ::= bitstring, fraction.

fraction ::= [].

fraction ::= ".", bitstring.

This grammar may be read declaratively as if it were a context free grammar. A *number,* for example, is a *bitstring* followed by a *fraction,* and a *fraction* is the terminal symbol "." followed by a *bitstring,* and so on. Procedurally, this grammar corresponds to a set of Horn clauses, much as do DCG rules. Here are the Horn clauses corresponding to Example 1:

```
bit(node(bit,[[0]],'[]'),Start,End) :-
   c(Start,'0',End).
bit(node(bit,[[1]],'[]'),Start,End) :-
   c(Start,'1',End).

bitstring(node(bitstring,'[]','[]'),Start,Start).
bitstring(node(bitstring,[BitTree,BitstringTree],'[]'),Start,End) :-
   bit(BitTree,Start,X),
   bitstring(BitstringTree,X,End).

number(node(number,[BitstringTree,FractionTree],'[]'),Start,End) :-
   bitstring(BitstringTree,Start,X),
   fraction(FractionTree,X,End).
```

```
fraction(node(fraction,'[]','[]'),Start,Start).
fraction(node(fraction,[['.'],BitstringTree],'[]'),Start,End) :-
    c(Start,'.',X),
    bitstring(BitstringTree,X,End).
```

They differ from the Prolog clauses representing DCG rules in that they have
three extra arguments attached to nonterminal symbols. The last two serve to
represent the difference list of characters being analyzed or synthesized (and are
used as in the Prolog clauses representing DCG rules). The third is used to
automatically construct the derivation tree of a successful parse. Each node of the
derivation tree is represented by a term of the following kind:

node(Nonterminal, List_of_subtrees, Semantics)

where *Nonterminal* is the name of the nonterminal symbol labeling the node in
the derivation tree, *List_of_subtrees* is a list of all the subtrees immediately
dependent from the node, and *Semantics* is a representation of the semantic rules
used to compute properties for that node. In Example 1, no semantic rules were
specified, so *Semantics* is always *[]*. A sample query using these rules is:

    :- number(Tree, "101.01", []), pretty(Tree).

with the following result:

```
number
  bitstring
    bit
      [1]
    bitstring
      bit
        [0]
      bitstring
        bit
          [1]
        bitstring
          []
  fraction
    [.]
    bitstring
      bit
        [0]
      bitstring
        bit
          [1]
        bitstring
          []
```

(*pretty* is the name of the predicate which prettyprints the derivation tree using indentation to indicate levels in the tree.)

Exercises.

1. Define *prettyprint*.

2. The grammar of Example 1 can be run "backwards," i.e., to generate strings of the language and corresponding trees. If Prolog's selection rule is used, however, not all strings and trees are generated. Explain. □

We will now add semantics to the grammar of Example 1. The semantics we shall add will treat the string of bits as representing a base 2 real number. We will consider each bit to have the value associated with its position:

```
number
  bitstring
    bit
     [1]        %value: 4
    bitstring
      bit
       [0]      %value: 0
      bitstring
        bit
         [1]    %value: 1
        bitstring
         []
  fraction
    [.]
    bitstring
      bit
       [0]      %value: 0
      bitstring
        bit
         [1]    %value: 0.25
        bitstring
         []
```

In order to compute these values we will have to do some tree traversals. Note that, given the tree structure, in order to assign the correct value to the leading bit we have to have an idea of how many bits there are in the string preceding the ".". If we calculate the length of the *bitstring* tree which is an immediate descendant of *number*, we can assign a "scale factor" to the leading bit (one less than the length). Two raised to the power of the scale power and multiplied by the bit gives the bit's value. The value of the next bit in the integer portion is found by decreasing the scale by 1, raising 2 to the scale factor, multiplying by the bit, and so on. In the fractional part we can begin with a scale factor of zero and decrease the scale factor for every bit to the right

```
number
  bitstring        %length: 3
    bit            %scale: 2; value: 4
    [1]
    bitstring      %length: 2
      bit          %scale: 1; value: 0
      [0]
      bitstring    %length: 1
        bit        %scale: 0; value: 1
        [1]
        bitstring %length: 0
        []
  fraction
    [.]
    bitstring
      bit
      bit          %scale: -1; value: 0
      [0]
      bitstring
        bit        %scale: -2; value: 0.25
        [1]
        bitstring
        []
```

The semantics we shall give will consist of rules attached to the derivation tree which will calculate two things: the length of a *bitstring* subtree, and the value of each bit.

In these rules we shall need some notation for specifying subtrees and for specifying the computation involving some rule labeling the subtree. In a grammatical rule we may associate a logical variable with a nonterminal symbol as follows:

nonterminal^^N

The logical variable N will eventually be instantiated to the subtree corresponding to *nonterminal*. In order to specify the computation of some semantic value X of a property *property* of the tree N we shall write:

N^property(X)

Some semantic properties might involve several values:

N^property2(X,Y)

The DCTG rule which specifies syntax must also specify semantics. This is done by including a rule for the computation of *property* to the right of the "<:>" symbol in the DCTG rule. Here is the grammar of our example with semantics as outlined above specified.

Example 2.
```
bit ::= "0"
<:>
bitval(0,_).

bit ::= "1"
<:>
bitval(V,Scale) ::- V is 2 ^ Scale.

bitstring ::= []
<:>
length(0),
 value(0,_).

bitstring ::= bit^^B, bitstring^^B1
<:>
(length(Length) ::- B1^^length(Length1),
            Length is Length1+1),
(value(Value,ScaleB) ::-
            B^^bitval(VB,ScaleB),
            S1 is ScaleB-1,
            B1^^value(V1,S1),
            Value is VB+V1).


number ::= bitstring^^B, fraction^^F
<:>
value(V) ::- B^^length(Length),
        S is Length-1,
        B^^value(VB,S),
        F^^fractional_value(VF),
        V is VB+VF.

fraction ::= ".", bitstring^^B
<:>
fractional_value(V) ::- S is - 1,B^^value(V,S).

fraction ::= []
<:>
fractional_value(0).
```

```
knuth(Source) :-
    number(Tree,Source,[]),
    pretty(Tree),nl,
    Tree^^value(Number),
    write(Number),
    nl.
```

In the DCTG rule for number, the subtrees for *bitstring* and *fraction* are named *B* and *F*, respectively. There is a single semantic property of *number* which is its *value*. The rule for computing *value* is read like a Horn clause. To the right of the "::-" symbol is a set of goals which must be executed in order to compute the value *V*. These may be read:

1. Compute the *length* of *B* in *Length*.
2. The scale *S* is 1 less than *Length*.
3. Compute the *value, VB* of *B* using *S*.
4. Compute the *fractional_value* of *F* in *VF*.
5. *V* is *VB+VF*.

The DCTG rules for *bitstring* have two semantic definitions, one for *length,* and one for computing *value.* The DCTG rules for *bit* have the semantic specification of *bitval* which defines how much each bit contributes to the number as a whole. The rules for *fraction* specify how the fractional value is added in to the number.

The controlling predicate *knuth* should be noted. After parsing the input string *Source* and obtaining the parse tree *Tree,* the semantic property *value* of *Tree* is obtained by traversing the tree. The semantic rules attached to each node constitute a local Horn clause data base for computations at that node. Selecting the appropriate rule, unifying variables in the head of the rule with the term being evaluated, and evaluation of subgoals proceeds as in ordinary logic programming. (See appendix II, 3.1 for a specification of the semantic interpreter).

The reader should be aware of some terminology used by Knuth and subsequently by people who work with attribute grammars. Knuth classified semantic properties either as *synthesized* or *inherited.* Synthesized attributes of a node are those whose values are computed entirely in terms of attributes of subtrees of the node; inherited attributes are those which are computed in terms of values which may come down from nodes higher in the tree or from sibling nodes. In our example, *length* is a synthesized attribute but *value* and *scale* are inherited because they each depend on the value of *length.* Knuth made the classification for a practical purpose. The attribute grammar writer might define a set of properties and a set of functions for computing these properties and introduce circularities into the computation of the semantic attributes. Knuth provides an algorithm for detecting such circularities by forming directed graphs in which the direction of arcs is determined by whether a property is inherited or synthesized. A cycle in such a graph is equivalent to a circularity in the definition of semantic attributes. In DCTGs, however, the power of the logical variable blurs the distinction of how and when a

property is calculated or instantiated, and so we have not insisted on any nota-
tional distinction between inherited and synthesized semantic attributes. If a
DCTG writer were to introduce a circularity in the definition of semantics, however, this would be equivalent to writing a logic program in which the occurs
check *must* be done.

Example 3.

In this example we shall define a subset of English sentences and translate them
into a logical form suggestive of Montague semantics. The translation will be
accomplished by specifying DCTG rules which, after parsing a sentence, traverse
a derivation tree to generate the logical form. Extension of this subset to a more
realistic subset of English can be left as a challenge to the reader. The logical
form used here is essentially the same as used earlier in the book. We are here
showing how logical form may be specified in the DCTG formalism.

Here is the underlying context free grammar which we shall develop.

sentence ::= noun_phrase, verb_phrase.

noun_phrase ::= determiner, noun, rel_clause.

noun_phrase ::= name.

verb_phrase ::= verb, noun_phrase.

verb_phrase ::= verb.

rel_clause ::= [that], verb_phrase.

rel_clause ::= [].

determiner ::= [every].

determiner ::= [a].

noun ::= [man].

noun ::= [woman].

noun ::= [cat].

name ::= [john].

name ::= [mary].

verb ::= [loves].

```
verb ::= [lives].

sentence(Source) :-
   sentence(T,Source,[]),
   pretty(T).
```

This grammar accepts sentences such as "Every cat loves a woman," but also such "sentences" as "Every cat lives a woman"; and, of course, no translation into logical form is possible with this grammar. We shall add semantics to this grammar to constrain accepted sentences to a more correct subset of English and to perform a translation to a suitable logical form. Here is sample output consisting of the parse tree and logical form for the sentence "Every cat loves a woman":

```
sentence
  noun_phrase
    determiner
     [every]
    noun
     [cat]
    rel_clause
     []
  verb_phrase
    verb
     [loves]
    noun_phrase
      determiner
       [a]
      noun
       [woman]
      rel_clause
       []
```

all(X,=>(cat(X),exists(Y,&(woman(Y),loves(X,Y)))))

This may be read as "For all X, if X is a cat, then there exists a Y such that Y is a woman and X loves Y."

Here is the full DCTG for our linguistic application. An explanation of the semantic rules follows the grammar.

```
sentence ::= noun_phrase^^N, verb_phrase^^V, !, { agreement(N,V) }
<:>
logic(P) ::- N^^logic(X,P1,P),
             V^^logic(X,P1).


noun_phrase ::= determiner^^D, noun^^N, rel_clause^^R
```

```
<:>
(agree(Num) ::- N^^agree(Num),
          D^^agree(Num),
          R^^agree(Num)),
(logic(X,P1,P) ::- D^^logic(X,P2,P1,P),
          N^^logic(X,P3),
          R^^logic(X,P3,P2)).

noun_phrase ::= name^^N
<:>
agree(singular),
(logic(X,P,P) ::- N^^logic(X)).

verb_phrase ::= verb^^V, { transitive(V) }, noun_phrase^^N
<:>
(agree(Num) ::- V^^agree(Num)),
(logic(X,P) ::- V^^logic(transitive,X,Y,P1),
          N^^logic(Y,P1,P)).

verb_phrase ::= verb^^V, { intransitive(V) }
<:>
(agree(Num) ::- V^^agree(Num)),
(logic(X,P) ::- V^^logic(intransitive,X,P)).

rel_clause ::= [that], verb_phrase^^V
<:>
(agree(Num) ::- V^^agree(Num)),
(logic(X,P1,&(P1,P2)) ::- V^^logic(X,P2)).

rel_clause ::= []
<:>
agree(Num),
logic(X,P,P).

determiner ::= [every]
<:>
agree(singular),
logic(X,P1,P2,all(X,=>(P1,P2))).

determiner ::= [a]
<:>
agree(singular),
logic(X,P1,P2,exists(X,&(P1,P2))).

noun ::= [man]
<:>
```

```
agree(singular),
logic(X,man(X)).

noun ::= [woman]
<:>
agree(singular),
logic(X,woman(X)).

noun ::= [cat]
<:>
agree(singular),
logic(X,cat(X)).

name ::= [john]
<:>
logic(john).

name ::= [mary]
<:>
logic(mary).

verb ::= [loves]
<:>
agree(singular),
logic(transitive,X,Y,loves(X,Y)),
logic(intransitive,X,loves(X)).

verb ::= [lives]
<:>
agree(singular),
logic(intransitive,X,lives(X)).

agreement(N,V) :-
    N^^agree(Num),
    V^^agree(Num).

transitive(V) :-
    V^^logic(transitive,_,_,_).

intransitive(V) :-
    V^^logic(intransitive,_,_).

sentence(Source) :-
    sentence(T,Source,[]),
    pretty(T),
    T^^logic(Proposition),
```

write(Proposition),nl.

Some further properties of the DCTG notation are illustrated in the right-hand side of the syntactic portion of the rule for *sentence* :

noun_phrase^^N, verb_phrase^^V, !, { agreement(N,V) }

DCTG rules allow the incorporation (in their Prolog setting) of the cut "!" for control. The "!" does not contribute to the parse tree, of course. Also, DCTG rules permit the inclusion within braces "{" and "}" of logic programming predicates. In the transformation of DCTG rules to logic program clauses, anything enclosed within the braces remains unchanged. These predicates may be thought of as expressing constraints on what is accepted by the grammar rule beyond what is defined by other grammar rules. In this case, the predicate *agreement* verifies that the noun phrase and verb phrase are in numerical agreement. Here is the translation of the rule for *sentence* and the definition of *agreement* :

```
sentence(node(sentence,[N,V],[(logic(P)::-
N^^logic(X,X1,P)','V^^logic(X,X1))]),Start,End) :-
   noun_phrase(N,Start,Mid),
   verb_phrase(V,Mid,End),
   !,
   agreement(N,V).

agreement(N,V) :-
   N^^agree(Num),
   V^^agree(Num).
```

Note that in the definition of *agreement,* the subtrees for the noun phrase and the verb phrase are traversed from within the DCTG rule itself in order to specify a constraint. The *agree* attribute of the noun phrase is used in this case to establish what the number of the noun phrase is (singular or plural), and this number is then used to make sure that the verb phrase has that same number. The arguments to *agreement* must, of course, be derivation trees with *agree* semantic attributes.

Constraints are also used to establish, in the rules for *verb_phrase* whether a verb is transitive or intransitive. The predicates that enforce the constraints check the subtree for *verb* to see whether the verb has an appropriate semantic property of *transitive* or *intransitive*.

In describing the semantic rules for *logic* which construct a logical representation of an acceptable sentence, it may be helpful to begin by describing what each lexical item contributes to the entire logical form. Names like "john" and "mary" are taken to be constants with respect to logical form.

Nouns like "cat" are taken to be one-place predicates of an existentially or universally quantified variable: *cat* (X). The variable is "generated" by a determiner.

The determiner "every" results in the generation of a logical component of the form:

all(X,=>(P1,P2))

which is read as "for all X, if P1, then P2" where P1 and P2 will be predications of X determined by other parts of the sentence. (Here reliance is placed on the power of the logical variable to eventually instantiate P1 and P2.) The determiner "a" results in the generation of the following logical component:

exists(X,&(P1,P2))

which is read as "there exists an X such that P1 and P2," where P1 and P2 are predications of X determined by other parts of the sentence.

Intransitive verbs are one place predicates. When the word "loves" is used as an intransitive verb, as in the sentence "John loves," the following logical form results:

loves(john)

Transitive verbs are two place predicates. When "loves" is used transitively as in the sentence "John loves mary," the following logical form results:

loves(john,mary)

The first argument is the subject, the second is the direct object. (A more extensive analysis would have to deal with indirect objects and adverbs governing the mode of action.)

The logical component of a nonempty relative clause is of the form:

&(P1,P2)

where P1 is a predication of the noun which the relative clause qualifies, and P2 is a predication introduced by the verb phrase within the relative clause.

With this notion of logical form in mind, the semantics which specify its construction may be followed. The main rule for sentence initiates the semantic traversal of the subtrees for *noun_phrase* and *verb_phrase* using the semantic rule:

logic(P) ::- N^^logic(X,P1,P),V^^logic(X,P1).

The construction of the logical form in P involves using a part of the logical form, that for the *verb_phrase,* which depends on information in the latter part of the sentence. Here, the logical variable is essential as a place marker. In following the construction of the logical form for the *noun_phrase,* one of the two following semantic rules are used, depending on whether the *noun_phrase* is a *name* or not:

logic(X,P,P) ::- N^^logic(X).

logic(X,P1,P) ::- D^^logic(X,P2,P1,P),
            N^^logic(X,P3),
            R^^logic(X,P3,P2).

If the *noun_phrase* is a *name*, the logical form of the nonterminal *name*, a constant, is found in the logic attribute of the name and X is instantiated to that constant. P is the logical form determined then entirely by the *verb_phrase* of the main sentence, hence its appearance twice in the rule. Otherwise, X is a variable which is quantified by the *determiner*, universally for "every," existentially for "a". In the second semantic rule, *P1* is eventually instantiated by the *verb_phrase* of the main sentence. The semantics for the embedded *noun* are determined in *P3* and may be further qualified by the semantics of a nonempty *relative_clause*:

logic(X,P1,&(P1,P2)) ::- V^^logic(X,P2).

P1 is the logical form for the *noun*, and *P2* is the additional component of logical form specified by the *verb_phrase*, determined by traversing the V tree and evaluating the logic attribute of V.

The logical form component for a *verb_phrase* depends on whether the verb is transitive or intransitive. If it is transitive, the rule is:

logic(X,P) ::- V^^logic(transitive,X,Y,P1),
            N^^logic(Y,P1,P).

X is the variable representing the subject; Y is a logical variable introduced to represent the object of the verb. The *noun_phrase* specifies the logical form for the object Y as detailed above. The first argument in the logic attribute for the verb is the term *transitive*, requiring that the lexical entry for the verb be marked with *transitive*.

If the verb is intransitive, the rule for specifying the logical form is simpler:

logic(X,P) ::- V^^logic(intransitive,X,P).

Here the predication P is made about the logical variable X which represents the subject of the sentence or sentence fragment.

Exercises

1. What is the logical form determined by the semantics of the Definite Clause Translation Grammar for the sentence: "Every cat loves a woman that loves a cat that loves a woman"?

2. Extend the grammar and semantics of logical form. What, for example, would be necessary in order to derive the logical form for the sentences "Cats love mary," "Cats love cats," "John loves every woman that loves cats"? □

2. A Compiler

2.1. Introduction

Some of the earliest applications of Definite Clause Grammars were to the specification and compilation of programming languages, indeed, even to the specification of Prolog compilers. The reason for this application is obvious: the syntax of programming languages can be specified to a large extent by context-free grammars, and the semantics or code generation phase of compilation can be specified as the result of some sort of derivation tree traversal. Given the DCG formalism, it was not long until it was applied to compiling technology with great success.

The compiler given below is an example of what is called syntax directed translation in compiling textbooks. The important difference between the compiler given here and those which might be given in conventional textbooks is that the logic grammar and logic programming specification provides both a declarative and procedural reading of the compiler. The specification is thus directly executable without the necessity of having to encode the specification in a lower level language (as is tacitly assumed in compilation textbooks). (The execution of the compiler, in view of the existence of Prolog compilers, need not be considered inefficient and merely a prototyped version of an eventual compiler coded in a conventional programming language.) If in general one thinks of a logic programming specification as being an executable program under its procedural reading, then with respect to logic grammars, the specification is the translator – a compiler or interpreter of the language in question.

We will demonstrate the way logic grammars can be used for compilation by specifying a compiler for a "toy" programming language using the DCTG formalism. Our language is toylike in the sense that although it is computationally complete (it could compute anything a Turing machine could) and it is small enough to describe in a section of a book. It is definitely not a language one would like to program in, but it will convey the basic ideas involved in developing full-scale logic grammar compilers. Our specification of this language will involve two grammars, in fact, one for specifying the lexical structure of the language ("lexical analysis") and the other for syntactic structure. Starting with a string of input characters, the output of the lexical structure grammar is a string of lexical tokens which becomes the input to the syntactic analysis grammar. The output of the syntactic structure grammar is a derivation tree from which an intermediate code generator produces an infix representation of the program; a code generator then translates the intermediate code to code for a very simple abstract machine; and, finally, an "assembler" translates the abstract machine code into something resembling a loadable form of machine code.

Here is an example of the compilation of a simple two-statement program:

read X; write X + 127
lexemes: [read,id(X),;,write,id(X),op(1,+),num(127)]


```
program
  statements
    statement
      tREAD
        [read]
      tIDENT
        [id(X)]
    st1
      tSEMICOLON
        [;]
      statements
        statement
          tWRITE
            [write]
          expression
            exp1
              primary
                tIDENT
                  [id(X)]
              exp2
                tOP
                  [op(1,+)]
                  exp1
                    primary
                      tCONSTANT
                        [num(127)]
                    exp2
                      []
                  exp2
                    []
        st1
          []
```

Code:
0: instr(read,5)
1: instr(load,5)
2: instr(addc,127)
3: instr(write,0)
4: instr(halt,0)
5: block(1)

At the end of this part we will direct the reader to various applications of logic grammars to the specification of programming languages and their compilers.

2.2. Lexical Analysis

The task of the lexical grammar is to specify how characters are grouped into meaningful tokens for syntactic analysis. These tokens include identifiers, integers, operator symbols, reserved words and punctuation marks. The translation of characters to tokens is essentially a gathering of characters into lists and the conversion of such lists to atoms. Since the specification of the lexical structure and translation to tokens is very simple, however, we shall write the lexical DCTG very much as if we were writing a Definite Clause Grammar. The DCTG notation is a generalization of the DCG notation, and grammars which look exactly like DCGs may be written in the DCTG notation, i.e., with arguments attached to non-terminal symbols. This aspect of the DCTG notation is particularly useful when the translation one wishes to specify is not complicated, can be accomplished in one *simple* pass, requires just one argument, and will not be expanded at some future time into something more elaborate.

The task of lexical analysis is to group characters into meaningful tokens while at the same time removing characters such as spaces, carriage returns, line feeds and comments which do not contribute to the meaning of a program. Also, it will be necessary to check whether some token that looks like an identifier has some reserved meaning. Here then is the top level of the grammar specifying the lexical structure of our simple language.

```
lexemes(L) ::= spaces, lexeme_list(L).
lexeme_list([L|Ls]) ::= lexeme(L), !, spaces, lexeme_list(Ls).
lexeme_list([]) ::= [].
```

```
lexeme(Token) ::=
    word(W) , { id_token(W,Token) }.
lexeme(Con)   ::= constant(Con).
lexeme(Punct) ::= punctuation(Punct) .
lexeme(op(Binding,Op)) ::= op(Binding,Op) .
lexeme(relop(Rel)) ::= relop(Rel).
```

The nonterminal *lexemes* forms a list of lexemes. A *lexeme* is either an *id_token*, i.e., a reserved word or an identifier, a constant, a punctuation symbol, an operator symbol (*op*), or a relational operator symbol (*relop*). A cut is used in the first rule for *lexeme_list* to indicate that a lexeme is deterministically defined.

Here are the rules relevant to specifying an *id_token*:

```
word([L|Ls]) ::= letter(L) , lords(Ls).
```

```
letter(L) ::= [L] , { is_letter(L) }.
```

```
is_letter(L) :- L>96, L<123. /* a-z */
is_letter(L) :- L>64, L<91. /* A-Z */

digit(D) ::= [D] , { is_digit(D) }.

is_digit(D) :- D>47, D<58. /* 0-9 */

lords([L|Ls]) ::= letter(L) , lords(Ls).
lords([L|Ls]) ::= digit(L) , lords(Ls).
lords([])     ::= [].
id_token(W,Token) :- name(X,W), token(X,Token).

token(X,Token) :- reserved(X,Token) , !.
token(X,id(X)).

reserved(div,op(2,intdiv)).
reserved(mod,op(2,mod)).
reserved(if,if).
reserved(then,then).
reserved(else,else).
reserved(while,while).
reserved(do,do).
reserved(read,read).
reserved(write,write).

spaces ::= space, spaces.
spaces ::= [].

space ::= " ".
space ::= [10].
```

This portion of the grammar should require little comment. The list of reserved words and symbols is specified in the unit clauses for *reserved,* with the second argument specifying the token which is represented by the atom in the first argument. Some reserved words such as *while* are represented by identical tokens; others, such as *div* undergo a transformation into a more complicated term. Predicates that specify what a letter or digit is are included and are obviously implementation dependent. In this case, the ASCII character set is used. The specification of *space* may be augmented by other character strings which can be ignored. This, too, is implementation dependent. The extra-logical predicate *name* converts a list of characters to an atom.

Following are the rules specifying a constant. In this simple language there is only one kind of constant, an integer constant. In more complex languages other constants, such as floating point numbers, for example, would have to be defined. The rule for *num* makes use of the above-mentioned extra-logical predicate *name* to convert a string of digits to an integer atom.

```
constant(C)  ::= num(C).

num(num(N)) ::= number(Number) , { name(N,Number) }.

number([D|Ds]) ::= digit(D) , digits(Ds).

digits([D|Ds]) ::= digit(D) , digits(Ds).
digits([])   ::= [].
```

Here are the definitions of punctuation characters for this simple language:

```
punctuation(lparen)      ::= "(" .
punctuation(rparen)      ::= ")" .
punctuation(':=')        ::= ":=".
punctuation(';')         ::= ";" .
```

The token corresponding to the character string specified by the right hand side of the grammar rule is specified by the argument attached to the nonterminal *punctuation*. Again, a more complicated language would have a larger set of punctuation marks.

Here are the allowed operators of our language. The integer argument in the definitions of *op* is used in syntactic analysis to determine the binding power of operators and the structure of expressions (see below). Notice that the character string for the *lt* token is a substring of the character string for the *le* token. Ordering of clauses is used to search for the longest alternative first.

```
op(1,'+') ::= "+" .
op(1,'-') ::= "-" .
op(2,'*') ::= "*" .
op(2,'/') ::= "/" .

relop(le) ::= "<=".
relop(lt) ::= "<" .
relop(ge) ::= ">=".
relop(gt) ::= ">" .
relop(ne) ::= "~=".
relop(eq) ::= "=" .
```

2.3. Between Lexical and Syntactic Analysis

The terminal symbols for syntactic analysis are the tokens resulting from lexical analysis. We could write rules for syntactic analysis by specifying the terminal symbols directly, as in:

```
primary ::= ['('], expression^^E, [')']
<:>
prefix(X) ::- E^prefix(X).
```

but instead we shall define a set of grammatical rules specifying the terminal symbols of our grammar for syntactic analysis. This gives a slightly more modular formulation in that a change in lexical structure could be made to a single rule rather than to all occurrences of a terminal symbol in the grammar for syntactic analysis. Also, as we shall see, some lexical symbols such as identifiers and constants carry more information than others. Our rules for synactic analysis will by convention specify terminal symbols by functors beginning with a lowercase "t":

```
primary ::= tLPAREN, expression^^E, tRPAREN
<:>
prefix(X) ::- E^prefix(X).
```

The following grammar rules constitute the interface between lexical and syntactic analysis. The semantics specified by *sem* in some of the rules are used in the code generation step to be explained below, and represent instructions in an abstract machine.

```
tLPAREN        ::= [lparen].
tRPAREN        ::= [rparen].
tASSIGN        ::= [':='].
tIF        ::= [if].
tTHEN          ::= [then].
tELSE       ::= [else].
tWHILE         ::= [while].
tDO         ::= [do].
tREAD        ::= [read].
tWRITE        ::= [write].
tSEMICOLON     ::= [';'].

tIDENT         ::= [id(Id)]<:>prefix(Id).

tCONSTANT      ::= [num(C)]<:>prefix(C).

tOP(1)         ::= [op(1,'+')]<:>prefix(add).
tOP(1)         ::= [op(1,'-')]<:>prefix(sub).

tOP(2)         ::= [op(2,'*')]<:>prefix(mult).
tOP(2)         ::= [op(2,'/')]<:>prefix(div).
tOP(2)         ::= [op(2,intdiv)]<:>prefix(intdivide).
tOP(2)         ::= [op(2,mod)]<:>prefix(modulus).

op_com         ::= [op(lt)]<:>prefix('<').
op_com         ::= [op(le)]<:>prefix('<=').
op_com         ::= [op(gt)]<:>prefix('>').
op_com         ::= [op(ge)]<:>prefix('>=').
op_com         ::= [op(eq)]<:>prefix('=').
```

op_com   ::= [op(ne)]<:>prefix('~=').

2.4. Syntactic Analysis

Here follows the syntactic portion of the DCTG for our simple programming language. (The semantic rules are discussed in the next section.) It is statement-oriented, and the nonterminal *statement* defines each of the different kinds of statement, namely an assignment statement, a while statement, an if statement, a read statement, a write statement, and a grouping of statements within parentheses. The allowable expressions are simple, consisting only of integer expressions using the small set of operators provided. Boolean expressions are modeled in the definition of the nonterminal *test* by a comparison of two integer expressions. A program consists of a sequence of statements.

The grammar rules for *expression* make use of a scheme whereby every arithmetic operator has a positive number attached to it specifying its binding power, with larger numbers indicating greater binding power. In parsing an expression, the nonterminal *exp1* begins parsing looking for operators with a binding power of 0. An *exp1* consists of a *primary* followed by an *exp2* which carries along with it the current binding power. In the first rule for *exp2*, if an operator is found with a greater binding power than *exp2* is currently expecting, *exp1* is used to gather up all parts of the expression involving operators with that higher binding power. Parsing then resumes with *exp2* looking for any remaining occurrences of operators with the original binding power. If no operator of greater binding power is detected by *exp2*, then the second rule for *exp2* is immediately satisfied by detecting the empty string. The reader should, if there is any confusion, parse by hand an expression such as $A*B*(1+C)+D$.

One further aspect of these rules might require some comment. In a purely logical grammar one could use left recursive rules with complete abandon. If, however, one has to rely on an implementation of logic programming which pursues a strategy of top-down, left-to-right evaluation, left recursive rules are forbidden: an infinite loop would simply be the result of attempting analysis or synthesis. There are techniques for removing left recursion and one of them is illustrated in the rules for *sum* and *product*.

    program ::= statements.

    statements ::= statement, st1.

    st1 ::= tSEMICOLON, statements.
    st1 ::= [].

    statement ::= tIDENT, tASSIGN, expression.
    statement ::= tWHILE, test, tDO, statement.
    statement ::= tIF, test, tTHEN, statement, tELSE, statement.
    statement ::= tREAD, tIDENT.
    statement ::= tWRITE, expression.
    statement ::= tLPAREN, statements, tRPAREN.

test ::= expression, op_com, expression.

expression ::= exp1(0).

exp1(Binding) ::= primary, exp2(Binding).

exp2(Binding) ::= tOP(Q), { Binding < Q },
          exp1(Q), exp2(Binding).

exp2(_) ::= [].

primary ::= tCONSTANT.
primary ::= tIDENT.
primary ::= tLPAREN, expression, tRPAREN.

An alternative strategy in parsing expressions is to use rules which are stratified in the sense that separate rules are used to detect subexpressions involving operators of different binding power. Here, for example, are alternate rules which do not make use of a notion of explicit binding power:

expression ::= sum.

sum ::= product, rest_of_sum.

product ::= primary, rest_of_product.

primary ::= tCONSTANT.
primary ::= tIDENT.
primary ::= tLPAREN, expression, tRPAREN.

rest_of_sum ::= op_add, product, rest_of_sum.
rest_of_sum ::= [].

rest_of_product ::= op_mul, primary, rest_of_product.
rest_of_product ::= [].

A *sum* is defined as a *product* followed by the nonterminal *rest_of_sum*; *rest_of_sum* may be empty, or it consists of an *op_add*, i.e., an addition or subtraction operator, followed by another *product* and another *rest_of_sum*. This comment about left recursion also explains why there are nonterminals *exp1* and *exp2*.

2.5. Code Generation

We present below the DCTG for our simple programming language, and we include all the annotations in the syntactic portions of the rules for naming trees, and also all the semantic rules which do most of the code generation. This code generation scheme follows that of Warren (1977), but of course, remodeled for the DCTG formalism.

2.5.1. The Target Machine

The code that is generated is for a simple single address computer with a single accumulator. The instructions for arithmetic are:

    add
    sub
    mult
    div
    modulus
    intdivide

    addc
    subc
    multc
    modulusc
    intdividec

where the instructions ending with the character ''c'' are used when the operand is a literal and can be packed into the instruction.

Data transfer instructions are:

    load
    loadc
    store

and these effect transfers between memory and the accumulator.

Instructions for jumping between points of the program follow. The conditional branch instructions test the contents of the accumulator, comparing the value therein to zero.

    jump

    jumpne
    jumpge
    jumple
    jumpeq
    jumplt
    jumpgt

For input and output there are two instructions:

    read
    write

There is an instruction for halting the machine:

halt

and pseudo-instructions for reserving a block of storage, and labeling a program point:

block
label

The instructions are represented in the DCTG by the two argument function term *instr*:

instr(op-code,operand)

In the arithmetic instructions, the operand is either the address of what is being added (subtracted, etc.) to the accumulator, or, in the case of literal operands, the value itself which is being added (subtracted, etc.) to the accumulator. Thus:

instr(add,200)

means add the contents of location 200 to the accumulator, but

instr(addc,200)

means add 200 to the accumulator. In the data transfer instructions, the operand is either an address or, in the case of the *loadc* instruction, a literal operand. In the branch instructions, the operand is the program point to which transfer of control is made. The operand of the *read* instruction is the location into which a value is read; for the *write* instruction, the contents of the accumulator are written out, but there is an operand which could be thought of as a device number. The operand for *halt* might be thought of as an error code.

The pseudo-instructions for reserving a block of storage, and for labeling program points are:

block(20)

label(20)

The first reserves a block of twenty memory locations, the second labels the program point as 20, counting from zero, with a "word" taking up a full memory location.

2.5.2. Code Generated for Statements

The DCTG rules defining statements have a single semantic rule attached to them for computing *code*. In the DCTG rule for *program*, *code* has two arguments: a logical variable *Dic* representing the dictionary or symbol table for the program, and a logical variable *Code* which is eventually instantiated to the code generated for a syntactically correct program. All code generated for the program makes use of the single dictionary *Dic* which is represented as a binary tree whose leaves are uninstantiated logical variables. If one looks up an entry in the dictionary, which has not yet been entered, the lookup predicate (see section 2.5.3 below), instantiates a logical variable to this entry and makes it a new branch node of the tree. This new instantiation is propagated by unification to every point where the common dictionary is used. This very nice feature of the logical variable provides a clean declarative and procedural interpretation to binary search in a dictionary of name and value pairs.

The semantic rule for the nonterminal *program* specifies that the dictionary and code for a program are the dictionary and code for the nonterminal *statements*. The semantic rule for *statements* specifies that if *SCode* is the code for the nonterminal *statement* generated using the dictionary *Dic,* and if *S1Code* is the code generated for the nonterminal *st1* (the remaining statements) using the same dictionary, then the code for *statements* using the dictionary *Dic* is the two element list: [*SCode,S1Code*]. The semantics of the first DCTG rule for·*st1* specify that the code for *st1* using the dictionary *Dic* is the code for the *statements* following the semicolon and using the same dictionary. If *st1* is empty, then its code is just the empty list, no matter which dictionary is used.

```
program ::= statements^^S
<:>
code(Dic,Code) ::- S^^code(Dic,Code).


statements ::= statement^^S, st1^^S1
<:>
code(Dic,[SCode,S1Code]) ::-
  S^^code(Dic,SCode),
  S1^^code(Dic,S1Code).

st1 ::= tSEMICOLON, !, statements^^S
<:>
code(Dic,Code) ::- S^^code(Dic,Code).

st1 ::= []
<:>
code(_,[]).
```

The Assignment Statement

statement ::= tIDENT^^Id, tASSIGN, expression^^E
<:>
code(Dic,[Exprcode,instr(store,Addr)]) ::-
 Id^^prefix(Identifier),
 lookup(Identifier,Dic,Addr),
 E^^code(Dic,Exprcode).

Given a dictionary *Dic,* the code for an assignment statement is a list consisting of the code for the expression, specified using the same dictionary, followed by a store instruction which stores the contents of the accumulator into *Addr,* the address of the identifier. The identifier is specified by the *prefix* semantic rule for the lexical-syntactic terminal *tIDENT.* The predicate *lookup* finds its address in the dictionary, making an entry for the identifier in the dictionary if it is not already present. Note that until allocation (see below), *Addr* will be an uninstantiated logical variable.

The While Statement

statement ::= tWHILE, test^^T, tDO, statement^^S
<:>
code(Dic,[label(L1),Testcode,Docode,instr(jump,L1),label(L2)]) ::-
 T^^code(Dic,L2,Testcode),
 S^^code(Dic,Docode).

For a given dictionary *Dic* which is used to generate the code for the test and the statement in the righthand side of the rule, *Testcode* and *Docode*, respectively, the code for the while statement is a list consisting of a label *L1,* the *Testcode,* the *Docode,* a jump instruction to the label *L1* (which is in front of the *Testcode*), followed by a label *L2* which labels the code following the while statement. *L2* labels the point to which control is transferred if the test at the head of the while statement fails. The "labels" *L1* and *L2* are uninstantiated variables until allocation and assembly (see below). At the end of the *Testcode,* there is a jump instruction to *L2* in case the test fails, hence, the second argument in *T^^code(Dic,L2,Testcode).*

The If Statement

statement ::= tIF, test^^T, tTHEN, statement^^S1, tELSE, statement^^S2
<:>
code(Dic,[Testcode,Thencode,
       instr(jump,L2),label(L1),Elsecode,label(L2)]) ::-
 T^^code(Dic,L1,Testcode),
 S1^^code(Dic,Thencode),
 S2^^code(Dic,Elsecode).

Given a dictionary *Dic,* the code for an if statement is a list consisting of the code for the test, *Testcode*, followed by the code for the statement following the *tTHEN,* followed by a jump instruction to label *L2* which labels the code following the if statement, followed by the label *L1,* the code for the statement following the *tELSE, Elsecode,* and the label *L2.* The code for the constituents of the if statement, *Testcode, Thencode,* and *Elsecode* are specified by recursive traversals of the parse tree, using the dictionary *Dic.* The second argument in *T^^code(Dic,L1,Testcode)* specifies the label to which control is transferred on failure of the test. All labels are uninstantiated until assembly and allocation.

The Read and Write Statements

```
statement ::= tREAD, tIDENT^^I
<:>
code(Dic,[instr(read,Addr)]) ::-
 I^prefix(Identifier),
 lookup(Identifier,Dic,Addr).
```

```
statement ::= tWRITE, expression^^E
<:>
code(Dic,[Exprcode,instr(write,0)]) ::-
 E^^code(Dic,Exprcode).
```

The code for a read statement is simply a read instruction: the address *Addr* of the specified identifier is looked up in the dictionary *Dic.* The code for a write statement is a list consisting of the code for the embedded expression, *Exprcode,* followed by a write instruction. The value left in the accumulator as a result of evaluating Exprcode is written on "device 0."

The Block

```
statement ::= tLPAREN, statements^^S, tRPAREN
<:>
code(Dic,Scode) ::- S^^code(Dic,Scode).
```

The code for the block (parenthesized statements) is simply the code for the parenthesized statements, using the same dictionary. (A more complex language would here require establishment of a new environment for definition of identifiers local to the block.)

The Comparision Test

```
test ::= expression^^E1, op_com^^O, expression^^E2
<:>
code(Dic,Label,[Exprcode,instr(Jumpif,Label)]) ::-
 E1^prefix(Arg1),
 E2^prefix(Arg2),
 O^prefix(Op),
 encode_prefix(expr(sub,Arg1,Arg2),0,Dic,Exprcode),
 unlessop(Op,Jumpif).
```

```
unlessop('=',jumpne).
unlessop('<',jumpge).
```

```
unlessop('>',jumple).
unlessop('~=',jumpeq).
unlessop('>=',jumplt).
unlessop('<=',jumpgt).
```

The code for a test is a list consisting of the code which computes the difference between the two embedded expressions in the test, followed by one of the conditional jump instructions. The target of the conditional jump is the location to which control is transferred if the specified comparision fails. The difference of the two expressions is formed in prefix notation which is convenient for generating code for expressions (see below).

2.5.3. Code for Expressions

In generating the code for expressions, it will be convenient to generate the code from Polish prefix notation, a representation different from the parse tree representation determined by the grammar for expressions. The reason for this is that the grammar is conditioned by the control strategy used in parsing (i.e., top-down, left-to-right recursive descent) which is reflected in the automatically produced parse tree. Thus, some of the structure of the parse tree does not represent the abstract structure of pure expressions consisting solely of operands and operators. For example, the expression $a+b$ is grammatically analyzed as:

```
expression
  exp1
    primary
      tIDENT
        [id(a)]
    exp2
      tOP
        [op(1,+)]
      exp1
        primary
          tIDENT
            [id(b)]
        exp2
        []
    exp2
    []
```

but, abstractly, all that is of interest are the subexpressions a and b and the operator "+".

Generating Prefix Expressions

The prefix form of an expression is generated by the *prefix* semantic attribute attached to the rules for expressions. For an *expression,* the prefix form is the same as the prefix form of an *exp1*, while for an *exp1*, the prefix form is specified by combining the prefix form of the *primary* with the prefix form of what follows in the *exp2*:

expression ::= exp1(0)^^E
<:>
(code(Dic,Code) ::- E^^prefix(Prefix),
  encode_prefix(Prefix,0,Dic,Code)),
(prefix(X) ::- E^^prefix(X)).

exp1(Binding) ::= primary^^P, exp2(Binding)^^E2
<:>
prefix(X) ::- P^^prefix(Primary),
        E2^^prefix(Primary,X).

In generating the prefix form of an *exp2*, *F* represents the prefix form of what is to the left of the current *exp2*. If there is an operator followed by an *exp1* and another *exp2*, the prefix form is generated by combining the *Operator, F,* and *F1,* the prefix form of *exp1,* as arguments to the functor *expr.* This, in turn, is passed to the subsequent *exp2* for completion of the prefix form of the rest of the expression. If, however, *exp2* is empty, then *F* is the prefix form of the entire *exp2*.

exp2(Binding) ::= tOP(Q)^^Op, { Binding < Q },
            exp1(Q)^^E1, exp2(Binding)^^E2
<:>
prefix(F,X) ::- Op^^prefix(Operator),
        E1^^prefix(F1),
        E2^^prefix(expr(Operator,F,F1),X).

exp2(_) ::= []
<:>
prefix(F,F).

The prefix form of a constant is of the form *num* (*X*) and that of an identifier is of the form *id* (*X*). The prefix form of a bracketed expression is the prefix form of the enclosed expression:

primary ::= tCONSTANT^^C
<:>
prefix(num(X)) ::- C^^prefix(X).

primary ::= tIDENT^^I
<:>
prefix(id(X)) ::- I^^prefix(X).

primary ::= tLPAREN, expression^^E, tRPAREN
<:>
prefix(X) ::- E^^prefix(X).

Code from Prefix Expressions

The semantic rules for *expression* specify that *Prefix* represents the prefix version of the expression represented by the parse tree *E,* and that the *Code* corresponding to *Prefix* is specified by evaluating the predicate:

encode_prefix(Prefix,0,Dic,Code)

The supplied dictionary *Dic,* initially an uninstantiated logical variable, is used as a symbol table. The integer 0 in *encode_prefix* represents the number of temporary locations previously used in code generation.

```
expression ::= exp1(0)^^E
<:>
(code(Dic,Code) ::- E^^prefix(Prefix),
  encode_prefix(Prefix,0,Dic,Code) ),
(prefix(X) ::- E^^prefix(X)).
```

The code that is generated for a number consists of the single instruction for loading a constant. Neither the dictionary nor the number of temporary variables is relevant in the generation of this instruction, hence the "_" as the second and third arguments of:

```
encode_prefix(num(C),_,_,instr(loadc,C)).
```

The code generated for an identifier consists of the single instruction for loading from the address associated in the dictionary with that variable:

```
encode_prefix(id(X),_,Dic,instr(load,Addr)) :-
  lookup(X,Dic,Addr).
```

The next two rules for *encode_prefix* deal with nontrivial prefix expressions, those which contain an operator. In the first rule, if *Expr2* is simple, the code generated is the list of the *Expr1code* followed by *Instruction,* where *Instruction* is generated as a result of a successful call of the predicate *simple.* An expression is simple if it is a number or an identifier, in which case a single instruction can be generated to perform the appropriate operation on the number or identifier and the contents of the accumulator. In the case of a number, the predicate *literalop* is used to generate an instruction that *contains* the number as a literal.

```
encode_prefix(expr(Op,Expr1,Expr2),N,Dic,
       [Expr1code,Instruction]) :-
  simple(Op,Expr2,Dic,Instruction),
  encode_prefix(Expr1,N,Dic,Expr1code).

simple(Op,num(C),_,instr(Opcode,C)) :-
  literalop(Op,Opcode).
```

```
simple(Op,id(X),Dic,instr(Op,Addr)) :-
   lookup(X,Dic,Addr).

literalop(add,addc).
literalop(sub,subc).
literalop(mult,multc).
literalop(div,divc).
literalop(modulus,modulusc).
literalop(intdivide,intdividec).
```

Finally, the following rule is used when *Expr2* is composite, i.e., not simple. The generated code consists of the code generated for computing *Expr2, Expr2code,* followed by an instruction to store the contents of the accumulator in the *N*th temporary location, followed by *Expr1code* and the instruction which performs the relevant operation on the contents of the accumulator (the value of *Expr1*) and the value saved in the *N*th temporary location. This is specified by verifying that *Expr2* is composite, looking up the address of *N* in the supplied dictionary, generating the code for *Expr2,* specifying *N1* as the successor to *N,* and generating the code for *Expr1* using *N1* in the second recursive call to *encode_prefix.*

```
encode_prefix(expr(Op,Expr1,Expr2),N,Dic,
         [Expr2code,instr(store,Addr),Expr1code,instr(Op,Addr)]) :-
   composite(Expr2),
   lookup(N,Dic,Addr),
   encode_prefix(Expr2,N,Dic,Expr2code),
   N1 is N + 1,
   encode_prefix(Expr1,N1,Dic,Expr1code).

composite(expr(_,_,_)).
```

Here is the specification of the predicate *lookup* , which uses binary search. It is presumed that some ordering is imposed on the atoms used as names. A dictionary is represented by the arity four functor *dic.* The first argument is a name or symbol in the dictionary, the second its value, the third a dictionary of names that occur before the current name in the ordering, and the fourth a dictionary of all succeeding names. Note that in the predicate lookup we make use of a binary tree whose leaves are uninstantiated logical variables. Looking up an atom in a tree that does not yet contain such an entry has the result of entering that atom in the binary tree dictionary in the proper place! Any such instantiation is propagated to all earlier references to that atom, e.g., in any generated code.

```
lookup(Name,dic(Name,Value,_,_),Value) :- !.

lookup(Name,dic(Name1,_,Before,_),Value) :-
   Name < Name1,
   lookup(Name,Before,Value).
```

```
lookup(Name,dic(Name1,_,_,After),Value) :-
  Name > Name1,
  lookup(Name,After,Value).
```

2.6. Assembly and Allocation

At this point the locations of variables in the virtual machine have not yet been determined, nor have addresses of branch targets. These quantities are still uninstantiated logical variables. For example, the code generated so far for the programming language statements:

read X; while X < 0 do read X

looks as follows:

```
0: instr(read,X)
label(L1):
1: instr(load,X)
2: instr(subc,0)
3: instr(jumpge,L2)
4: instr(read,X)
5: instr(jump,L1)
label(L2):
```

In order to substitute locations for the virtual machine's variables and branch targets we make use of the following predicates: *assemble* and *allocate*. The code generated so far consists of a list whose elements are either labels or instructions or sublists of generated codes. The first argument of *assemble* is such a list or an instruction or a label, the second and third arguments are integers representing the locations of the first and last instructions in the list. Assembly of a list of instructions consists of assembling the first element of the list, followed by assembly of the rest of the list. If the list is empty, assembly is complete.

```
assemble([Code|Codes],N0,N) :-
  assemble(Code,N0,N1),
  assemble(Codes,N1,N).
```

```
assemble([],N,N).
```

Assembly of an instruction merely increments the count of locations used to hold the program:

```
assemble(instr(_,_),N0,N) :-
  N is N0 + 1.
```

A label is assembled by assigning the location of the next instruction to the logical variable representing the label:

```
assemble(label(N),N,N).
```

In order to allocate virtual machine locations to the variables used in the generated code, the predicate *allocate* traverses the dictionary in inorder. Uninstantiated dictionaries are instantiated to the atom *void*:

```
allocate(void,N,N) :- !.
```

Nonempty dictionaries are traversed in inorder by means of the following recursive rule:

```
allocate(dic(Name,N1,Before,After),N0,N) :-
  allocate(Before,N0,N1),
  N2 is N1 + 1,
  allocate(After,N2,N).

compile(Source) :-
    nl,
    writestring(Source),nl,
    lexemes(Tokens,_,Source,[]),
    writel(['lexemes: ',Tokens,nl]),
    program(Tree,Tokens,[]),
    nl,pretty(Tree),nl,
    Tree^^code(Dic,Code),
    assemble(Code,0,N0),
    N1 is N0 + 1,
    allocate(Dic,N1,N),
    L is N - N1,
    writel(['Code: ',nl]),
    writecode([Code,instr(halt,0),block(L)],0,_).

writecode([],N,N) :- !.

writecode([C|Cs],N0,N1) :-
    writecode(C,N0,N),
    writecode(Cs,N,N1).


writecode(instr(Op,Oper),N,N1) :-
    writel([N,': ']),
    write(instr(Op,Oper)), nl,
    N1 is N + 1.

writecode(label(L),X,X) :-
    writel([label(L),': ',nl]).
```

```
writecode(block(L),X,X) :-
   writel([X,': ',block(L)]).


writel([nl|Xs]) :- !,
   nl,
   writel(Xs).

writel([X|Xs]) :-
   write(X),
   writel(Xs).

writel([]).
```

3. A Metagrammatical Extension of DCTG Notation

Although context-free rules can be used to describe much of programming
language syntax, there are some drawbacks to the notation which result in
inelegant repetition of similar rules. For example, very often one wishes to specify
that a certain nonterminal symbol x may occur optionally. This may be accomplished crudely by copying all rules in which x occurs with the difference that in
the copy, x does not occur. For example:

```
n ::= ...,y,x,z...
n ::= ...,y,z,...
```

Optionality of x may be accomplished somewhat more elegantly by introducing a
new nonterminal symbol a defined as follows:

```
a ::= x.
a ::= [].
```

and then n is defined by:

```
n ::= ...,y,a,z...
```

However, this requires the grammar-writer to add a new nonterminal and two new
rules for each such nonterminal. To remedy this situation, some sort of notation
such as $<x>$ might be introduced to indicate optionality thus:

```
n ::=...,y,<x>,z...
```

It is not only optionality, however, that is a convenient addition to context-free
notation. One might also want to indicate the occurrence of a grammatical symbol
one or more times, or zero or more times: in these cases a convenient notation
derives from the Kleene closure. x^* denotes a possibly empty sequence of xs,
while x^+ denotes a nonempty sequence of xs.

More generally, however, one might wish to indicate the occurrence of certain patterns of grammatical symbols without introducing more nonterminal symbols. In defining Algol 68, for example, a notation was used which permitted the specification of *packs*, which generalized parenthesized or bracketed constructs such as:

(expression)

begin statements **end**

[nonempty array index sequence]

(parameter list)

Another example arises from the fact that data structures may be specified by means of grammatical rules (see next section for more on this notion). Here is a DCTG rule for a *tree* with *identifiers* as leaves:

tree ::= identifier.
tree ::= "(", tree, ",", tree, ")".

If we also wanted to define a tree with *numbers* as leaves, we would have to introduce another nonterminal *ntree* and rules defining it:

ntree ::= number.
ntree ::= "(", ntree, ",", ntree, ")".

Here too we run into the problem of duplication of rules with essentially the same structure. Ideally, we would like to define a general *tree* parameterized by the kind of leaf it should have:

tree(Leaf) ::= Leaf.
tree(Leaf) ::= "(", tree (Leaf), ",", tree (Leaf), ")".

The compiler from Definite Clause Translation Grammar notation to Prolog (see appendix II) does not permit the easy specification of any of the constructs mentioned above. It is possible, however, to introduce a single metanonterminal symbol which permits the above mentioned extensions of grammatical notation, and many more besides. This nonterminal symbol acts as a grammatical metacall facility, akin to Prolog's metacall and is written *meta(x)* where x is some grammatical symbol, either a terminal or a nonterminal symbol. During execution (parsing or generation), occurrence of *meta(x)* in a grammatical rule results in the metacall of the symbol x: if x is a nonterminal symbol, the hidden extra arguments are attached to x and analysis or generation continues, either looking for or generating an occurrence of x.

This mechanism provides a means of metagrammatical abstraction, permitting many useful extensions of the basic, underlying DCTG notation. (Of course, the notion of metagrammatical abstraction may be adapted to other formalisms as well.) We can now easily specify the following rules for the optional occurrence of the symbol x:

    option(X) ::= meta(X)^^H
    <:>
    Sem ::- H^^Sem.

    option(X) ::= [].

In the first DCTG rule, the semantic portion guarantees the copying of the semantic rules (if any) from X to the rule in which the metacall occurs. We would use it as follows:

    n ::= ...,y,option(x),z...

The predicate *meta* now allows us to specify different kinds of trees concisely:

    tree(X)::= meta(X).

    tree(X)::= "(", tree(X), ",", tree(X), ")".

This metadefinition of *tree* now permits the following usage:

    n ::= ...,y,tree(x),z...

where x is the grammatical symbol which should constitute the leaves of our tree.

A sequence of zero or more occurrences of a grammatical symbol is defined by the following DCTG metarule:

    sequence(Thing) ::= meta(Thing)^^H, sequence(Thing)^^S
    <:>
    list([Head|Tail]) ::- H^^element(Head),S^^list(Tail).

    sequence(Thing) ::= []
    <:>
    list([]).

Here, we suppose that if a sequence of some nonterminal x is to be formed that x has a semantic attribute *element,* which may be collected into a sequence of xs by the semantic rule for *list.* The empty sequence generates the empty list of elements.

We now specify a lexical analyzer using DCTG rules with metacalls. Within that specification we define *numeral* and *identifier* using the definition of *sequence*:

numeral ::= digit^^D, sequence(digit)^^S
<:>
element(Num) ::-
    D^^element(Digit),S^^list(Digits),name(Num,[Digit|Digits]).

identifier ::= letter^^L, sequence(lord)^^S
<:>
element(Id) ::-
    L^^element(Letter),S^^list(Lords),name(Id,[Letter|Lords]).

The nonlogical predicate *name* constructs an atom from a list of either digits or alphanumeric characters. Both *numeral* and *identifier* have a semantic definition for *element* as numerals and identifiers may be lexemes in a list of lexemes.

We now introduce two new metarules, *list_of1([X,Y])* which generates a list of pairs of the grammatical symbols X and Y : $X,Y,X,Y,\cdots,X,Y$ where we suppose that the X symbols have a semantic attribute element, i.e., the Ys act as separators of the Xs; *list_of2([X,Y])* which generates a list of pairs of the grammatical symbols X and Y:$X,Y,X,Y,\cdots,X,Y$ where we suppose that the Y symbols have a semantic attribute element, i.e., the Xs act as separators of the Ys.

list_of1([X,Y]) ::= meta(X)^^H, meta(Y), list_of1([X,Y])^^S
<:>
list([Head|Tail]) ::- H^^element(Head),S^^list(Tail).

list_of1([_,_]) ::= []
<:>
list([]).

list_of2([X,Y]) ::= meta(X), meta(Y)^^H, list_of2([X,Y])^^S
<:>
list([Head|Tail]) ::- H^^element(Head),S^^list(Tail).

list_of2([_,_]) ::= []
<:>
list([]).

We also define:

alternating_list1([A,B]) ::= meta(A),list_of1([B,A])^^Bs
<:>
list(L) ::- Bs^^list(L).

alternating_list2([A,B]) ::= meta(A)^^FirstA,list_of2([B,A])^^As
<:>
list([L|Ls]) ::- FirstA^^element(L),As^^list(Ls).

In these constructs, the semantic attribute *list* gathers together the elements of interest in the respective lists.

We may then define *lexemes* using our metagrammatical abstractions as:

    lexemes ::= alternating_list1([spaces,lexeme])^^Lex
    <:>
    list(L) ::- Lex^^list(L).

where

    space ::= " ".

    spaces ::= sequence(space).

The remaining rules (with obvious omissions and possibilities for extension) for our lexical analyzer specified with metarules are:

    digit ::= [D], { is_digit(D) }<:>element(D).

    letter ::= [L], { is_letter(L) }<:>element(L).

    lord ::= letter^^L<:>element(Lord) ::- L^^element(Lord).

    lord ::= digit^^D<:>element(Lord) ::- D^^element(Lord).

     or([X|_]) ::= meta(X)^^H<:>Sem ::- H^^Sem.

    or([_|X]) ::= or(X)^^H
    <:>
    Sem ::- H^^Sem.

    lexeme ::= or([".", "(", ")", identifier,numeral])^^H
    <:>
    element(X) ::- H^^element(X).

The nonterminal *or* can be used to recognize any one of a list of grammatical symbols. Now this definition of lexical analysis using metarules appears to be somewhat more complicated than the lexical analyzer given earlier. However, we have introduced a notion of metagrammatical abstraction that is new, which permits useful extensions to our notation, and consequent compressions of rule definitions. For example, one of the metaconstructs introduced in the Algol 68 Report was for a generalized form of bracketing called a pack which in our notation is given by:

    pack([Left,X,Right]) ::=meta(Left), meta(X)^^H, meta(Right)

```
<:>
Sem ::- H^^Sem.
```

The semantics for *pack* are exactly those of the bracketed construct *X*. The examples discussed above can now be specified in DCTG rules by reference to:

```
pack(['('],expression,[')'])
```

```
pack([begin], statements, [end])
```

```
pack(['['], non_empty_sequence(array_index) , [']'])
```

```
pack(['(',param_list,')'])
```

Considering the last of these we can define *param_list* as:

```
param_list ::= alternating_list2([ identifier, ','])^^Param
<:>
list(L) ::- Param^^list(L).
```

and *parameter_pack* as:

```
parameter_pack ::= pack(['(',param_list,')'])^^Param
<:>
list(L) ::- Param^^list(L).
```

Thus, by writing a few metarules, grammatical abstraction makes it possible to replace repeated occurrences of similar grammatical constructions with concise definitions.

Exercises

1. Using the above-given metarules, and any necessary grammatical rules, define as concisely as possible the remaining pack constructs. □

Finally, here, with reference to appendix II defining the compiler from DCTG rules to Prolog, is one possible specification of the predicate *meta:*

```
meta([X],[X],Input,Output) :-c(Input,X,Output).
```

```
meta(X,Tree,Input,Output) :-
    X =.. XList,
    append(XList,[Tree,Input,Output],NewXList)
    NewX =.. NewXList,
    NewX.
```

The first clause for *meta* deals with recognition of a metaterminal symbol [*X*].
The first argument is the metasymbol, the second its representation in the derivation tree, and the third and fourth the difference lists representing the string being
analyzed. The second clause deals with recognition of a metanonterminal symbol
X. The symbol is transformed into a list *XList* and is appended to the list of arguments consisting of the tree and the difference list representing the string being
analyzed [*Tree,Input,Output*]. This list is then transformed into a predicate symbol
NewXList which is then called. If for example we had written
meta(nonterminal(A)), then

> X = nonterminal(A),
> XList = [nonterminal, A]
> NewXList = [nonterminal, A, Tree, Input, Output]
> NewX = nonterminal(A, Tree, Input, Output)

When this definition is used, calls to *meta* are interpreted at analysis or generation
time, that is, the metarules are constructed on the fly.

Exercises

1. Can the definition of *meta* be rewritten so as to generate new DCTG rules at
   compile time, thus generating Prolog clauses and avoiding interpreted calls?
   How is this done?

2. Investigate the use of *meta* for natural language applications. We might define,
   for example, a metaconjunction rule as follows:

> conjoin(X) ::= [either],meta(X)^^X1,[or],meta(X)^^X2
> <:>
> meaning(and(Y,Z)) ::- X1^^meaning(Y),X2^^meaning(Z).

Here the semantic portion's *meaning* of the conjunction is the constructed
symbol *and(Y,Z)* formed by composing the meanings of the conjoined components. The rule might be used in:

> n ::= ...,conjoin(verb_phrase),....

or:

> m ::= ...,conjoin(sentence),....

See Abramson (1988) for details on a solution to this exercise. □

4. Grammatical Data Typing

So far, we have focused on the use of logic grammars for the analysis and generation of strings of the language defined by a given grammar; the derivation tree,
representing the structure of sentences, has been viewed as a useful by-product of
the analysis or generation phase. In this section we shall shift our focus a bit and
consider the use of grammatical notation as a way of specifying data structures or
types. The basic idea is that data types can be considered to be unambiguous
context-free grammars: each nonterminal of a grammar represents a type whose
construction is specified by the right-hand side of one or more productions; furthermore, the nonterminals in the right-hand side of a production act as selectors for
decomposing elements of the type, essentially, derivation trees. The constructor
and selectors can then be used to define functions or relations over the type and
between types specified in this manner.

The grammatical notation we shall use for this is a simple extension of DCTG
notation. Given a production defining a nonterminal X in our extended DCTG
notation, the nonterminals in the right-hand side of the production will be treated
as semantic attributes of X. These attributes may be used either to decompose an
object of type X into its constituents or to compose an object of type X from an
appropriate set of components. These semantic attributes, in combination with
type-checking predicates, which can be generated automatically, may be used to
define relations over a type or between types. Automatically generated type checking predicates may be used either to verify that an object is of a certain type, or to
generate an object of a certain type. We shall proceed with several examples illus-
trating the concepts.

4.1. The Natural Numbers

The natural numbers are specified by the following grammar:

> zero:natural ::= ''0''.
> succ:natural ::= ''s('', natural, '')''.

Terminal symbols that are strings or lists of characters are enclosed within quotation marks. (Terminals may also be indicated as a list of nullary function symbols,
e.g., [*symbol*].) The first rule specifies that 0 is a *natural* (*natural* is the only
nonterminal of this grammar). It also specifies that the name of this production is
zero and that there exists a unary predicate *zero* (X) which is satisfied only if X is
a derivation tree whose leaves, read in order from left to right, are in the language
generated by this production. The second rule specifies that the other form of a
natural is $s($ followed by a *natural* followed by $)$, for example, $s(0)$ or $s(s(0))$.
The name of this production specifies a unary predicate *succ* (X) which is
satisfied if X is a derivation tree whose leaves read in order from left to right are in
the language generated by this grammar.

The relation *pred* specifies that N is the predecessor of X :

    pred(X,N) :-
            succ(X)^^[natural(N)].

Here, *succ* (*X*) specifies that *X* is of type *natural* and is generated by the production named   *succ,* i.e., *X* is a derivation tree for some sentence generated by the
grammar for   *natural,* and the first step in the derivation of the sentence uses the
production *succ.* Reading the leaves of *X* would therefore give us something of
the form *s* (*N*) where *N* is a *natural.* The notation introduced consists of an infix
operator ^^ from Definite Clause Translation Grammars and is read as "with subtrees such that," followed by a list of unary function symbols applied to logical
variables, in the example, [*natural* (*N*)]. The unary function symbols must be in
the set of nonterminals which appear in the right-hand side of the applicable production, here the production named *succ.* Each such unary function symbol (here
only *natural*) selects the relevant subderivation tree of the derivation tree named
in the predicate to the left of the ^^ operator, and instantiates its argument to it.
Thus, if the leaves of *X* read in order are *s*(, 0 and), then the natural number *N* is
a subtree of *X,* consisting of the leaf 0. Since *pred* is a relation, we are also
specifying that *X* is the successor of the natural number *N.* We can read
succ(X)^^[natural(N)]: for any natural number *X* not equal to zero, if *N* is the
predecessor of *X,* then the successor of *N* is *X.* In functional notation, we might
write that *X =succ* (*pred* (*X*)). (We are also being slightly informal here: we should
mention "the leaves of *X* read in order from left to right," etc. Since the grammars we are using for this application are unambiguous we can identify derivation
trees with the strings labeling their leaves read in the right order.)

The type *natural* contains all derivation trees generated by the grammar given
above. We specify this by the clauses:

    type(X,natural) :- zero(X).
    type(X,natural) :- succ(X).

This is read "*X* is of type *natural* if it satisfies *zero* or *succ*." *X* might be considered a grammatical variable of type   *natural,* i.e., *X* ranges over derivation
trees of sentences in the language generated by the grammar. The predicates
zero, succ and *type* can be generated automatically as the grammar is compiled
into Prolog clauses.

Information about each production for a type is also recorded in unit clauses of
attribute :

    attributes(succ,natural,[natural(X)]).
    attributes(zero,natural,[]).

The first argument is the name of a production, the second a type, and the third a
list of the applicable selectors. This could be used in static type checking.

Translating the Peano axioms for addition into our notation, we have:

    sum(X,Y,X) :-
            type(X,natural),

zero(Y).

sum(X,Y,S) :-
    type(X,natural),
    succ(Y)^^[natural(P)],
    succ(S)^^[natural(Q)],
    sum(X,P,Q).

If X and Y are instantiated in the second clause of *sum,* that is, if X and Y are derivation trees for natural numbers, then $succ\,(S)$^^[*natural*$\,(Q)$] specifies that S is a derivation tree from production *succ* with subderivation tree Q, Q a *natural*; Q is instantiated as a result of the recursive call of *sum.* The predicate *sum* in fact specifies a relation between three naturals. The reader may verify that if $S=s\,(s\,(s\,(0)))$ then *sum* may be used to find all natural numbers X and Y which add up to S. In this case note that *type(X,natural)* and $succ\,(Y)$^^[*natural*$\,(P)$] act as generators of X and Y rather than as type verifiers.

The two lines of the first clause for *sum* and the first three lines of the second clause are suggestive of the type declarations, with initialization, of some Von Neumann languages; the last line of the second clause is suggestive of the body of a block or procedure.

The nonterminal symbols of this extended grammatical notation corresponds to predicates (as in DCTG's) with three extra arguments representing the derivation tree and the two components of the difference list of terminal symbols. Thus, a query of the form

    :- natural(Tree,''s(s(0))'',[])

would succeed and instantiate *Tree.* The representation for tree nodes is similar to that used in ordinary DCTG notation, but modified to take into account the extra notation used here. See appendix II, section 3.3.

4.2. Lists

Here is a grammar which defines simple lists:

    nonempty:list ::= string, '','', list.
    empty:list ::= [].

A list is either *empty* or it consists of a *string* followed by a comma followed by a *list. string* is a primitive type and consists either of a sequence of numerical characters or of a letter followed by zero or more letters or digits. (The original characters of the sequence are converted to a nullary function symbol.) Thus, 123 and *i2* are strings. The following are, therefore, acceptable lists: *1,2,3,* and *abc,def,12,.* We shall not list here the clauses of *type* and *attribute* defined by this grammar.

We define the relation *append* between three arguments of type list:

```
append(E,X,X) :-
   empty(E),
   type(X,list).

append(X,Y,Z) :-
   nonempty(X)^^[string(A),list(XX)],
   type(Y,list),
   nonempty(Z)^^[string(A),list(ZZ)],
   append(XX,Y,ZZ).
```

The types which have been specified prevent this version of *append* from the poor behaviour of the usual Prolog *append*. Using that version of *append*, for example, one may *append* a list [*a,b,c*] to a term such as 4, which is not itself a list. This version of *append*, of course, may be used nondeterministically to generate all lists *X* and *Y* which when appended yield, for example, *abc,def,12,* .

We specify a predicate *length* between the types *list* and *natural* as follows:

```
length(L,Zero) :-
   zero(Zero),
   empty(L).

length(L,N) :-
   type(N,natural),
   nonempty(L)^^[list(L1)],
   length(L1,N1),
   one(One),
   sum(N1,One,N).
```

The predicate *one* specifies the successor of "0":

```
one(One) :-
   zero(Zero),
   succ(One)^^[natural(Zero)].
```

The specifications of the types of *N1* and *One* could be made explicit by adding *type(N1,natural)), type(One,natural)* to the definition of the second clause of length, but could also be inferred from the type requirements of *one* and *sum* by a static-type checker.

4.3. Trees

The following grammar specifies the type *tree*:

```
leaf:tree::=string.
branch:tree::= "(", left:tree, ",", right:tree, ")".
```

In the previous grammars we were able to use the names of the nonterminals in the right-hand side of a production as the selectors (decomposers) of the type defined by that production, making "puns," so to speak, with the names of the nonterminals in the right-hand side: context clarifies whether we are talking of a "selector" or a "type." We cannot do this here since there are two occurrences of the type *tree* in the right hand side of the branch production. These occurrences of *tree* are, however, labeled *left* and *right,* and these labels are the selectors of trees which are branches in the *deepreverse* predicate below which reverses a *tree* at all levels. (We need not label both uses of *tree* in the right-hand side: labeling one would remove the ambiguity of which subtree was meant). The clauses for *type* are not shown.

```
attributes(branch,tree,[right(R),left(L)]).
attributes(leaf,tree,[string(S)]).

deepreverse(X,X) :- leaf(X).
deepreverse(X,Y) :-
    branch(X)^^[left(Left),right(Right)],
    branch(Y)^^[left(RRight),right(RLeft)],
    deepreverse(Left,RLeft),
    deepreverse(Right,RRight).
```

4.4. Infix and Prefix Notation

In our final example we specify *infix* and *prefix* expressions and a predicate *convert* which allows one to *convert* between them. The predicate *convert* may be used, of course, in either direction and may be thought of as specifying a source-to-source translation of a simple kind, translating between expressions such as $a*(b+c)$ and $*a,+b,c$. As in the previous examples, the definition of *convert* is recursive and split into cases depending on the grammatical structure of its arguments. The predicates *type* and *attribute* generated for this grammar are not shown.

```
p1:infix ::= expression.
p2:expression ::= term, "+", expression.
p3:expression ::= term.
p4:term ::= primary, "*", term.
p5:term ::= primary.
p6:primary ::= string.
p7:primary ::= "(", expression, ")".
```

r1:prefix ::= "+", pre1:prefix, ",", pre2:prefix.
r2:prefix ::= "*", pre1:prefix, ",", pre2:prefix.
r3:prefix ::= string.

convert(In,Pre) :-
    p1(In)^^[expression(E)],
    type(Pre,prefix),
    convert(E,Pre).

convert(In,Pre) :-
    p2(In)^^[term(T),expression(E)],
    r1(Pre)^^[pre1(PT),pre2(PE)],
    convert(T,PT),
    convert(E,PE).

convert(In,Pre) :-
    p3(In)^^[term(T)],
    type(Pre,prefix),
    convert(T,Pre).

convert(In,Pre) :-
    p4(In)^^[primary(P),term(T)],
    r2(Pre)^^[pre1(PP),pre2(PT)],
    convert(P,PP),
    convert(T,PT).

convert(In,Pre) :-
    p5(In)^^[primary(P)],
    type(Pre,prefix),
    convert(P,Pre).

convert(In,Pre) :-
    p6(In)^^[string(S)],
    r3(Pre)^^[string(S)].

convert(In,Pre) :-
    p7(In)^^[expression(E)],
    type(Pre,prefix),
    convert(E,Pre).

4.5. Comments on Grammatical Typing

Traditionally, grammars have been viewed as devices for analyzing and generating terminal strings of a language. In this section, the focus has been on the derivation trees that the grammar can produce. The trees may be only *partially instantiated,* derivation trees of terminal strings being a special case of general derivation trees. Context-free grammars, suitably analyzed, provide all the necessary selectors and constructors for manipulating derivation trees of the grammars. The extension of DCTG notation for typing thus provides a very simple method of imposing an optional type discipline on logic programs.

These extended grammars can, of course, be used in the traditional manner to analyze or generate a terminal string. Once the string has been parsed, however, the interest generally lies in manipulation of the derivation tree to produce other members of the type, or to select components of the tree, etc. A typical query is:

```
:- natural(N1,"s(s(0))",[]),
     natural(N2,"s(s(s(0)))",[]),
     sum(N1,N2,N3).
```

where once parsing is complete, some arithmetic processing is done. The compilation of typing DCTG rules to Prolog clauses, in fact, forces a restriction on the use of typing DCTGs as parsers: Prolog's top-down left-to-right strategy rules out left recursive productions. (The restriction could be removed by using Earley's general context-free parsing algorithm; this would permit one to specify all data structures in their most "natural" form even if it meant using left recursive rules.) However, if one uses the generators and composers implicit in a grammar to form derivation trees without parsing terminal strings, then even left recursive rules could be used in a grammar. Consider the following (more abstract) definition of *tree* :

```
leaf:tree ::= string.
branch:tree ::= left:tree, right:tree.
```

A *branch* may be constructed as a result of the following query:

```
:- string(S,"abcd",[]),
     string(T,"1234",[]),
     leaf(L1)^^[string(S)],
     leaf(L2)^^[string(T)],
     branch(B)^^[left(L1),right(L2)].
```

without using the branch rule in parsing.

Note that data types that are not context-free may be specified by placing restrictions or constraints on derivation trees. For example:

```
is_p:prime ::= natural^^P, { is_prime(P) }.
```

specifies that a prime is a natural number P constrained by a call of a predicate *is_prime* (P) to be a prime number. In writing predicates over this type, it is allowed to include predicates in the specification lists operated on by $^{\wedge\wedge}$:

is_p(P)^^[natural(P), {is_prime(P)}].

This is read as "the implicit (generated) predicate *is_p* is true of *P* provided that *P* is a *natural* and that the additional constraint *is_p* is also satisfied by *P*."

Typing DCTGs provide a run-time type checking mechanism based on unification. This can prove expensive for large data structures. One way to remedy this would be to provide a syntactic sugaring of Prolog programs which would permit types to be specified and statically checked by a grammar based type checker. A well-typed Prolog program could then use a simpler and more efficient representation of types than the structure shown above, and potentially expensive unifications could be avoided: a program would work because it had been shown to be well-typed.

Chapter 10

Further Expressive Power–Discontinuous Grammars

1. The Discontinuous Grammar Family

Discontinuous grammars (DG)[1] were devised by V. Dahl in 1981, as a generalization of extraposition grammars. They are basically metamorphosis grammars with the added flexibility that unidentified strings of constituents can be referred to (usually through a pseudo-symbol skip(X), where X stands for the skipped substring), to be repositioned, copied, or deleted at any position. They do not need to obey, as do XG's, any nesting constraints.[2]

More formally, a discontinuous grammar is a quintuple (V_N, V_T, K, s, P), where V_T and V_N are the terminal and nonterminal vocabularies respectively, whose union is called V, $s \in V_N$ and is the starting symbol, K is the set of skip symbols, with K and V non-intersecting, and P is a set of productions of the form

$$nt, \alpha_0, skip(X_1), \alpha_1, \ldots, skip(X_n), \alpha_n \dashrightarrow$$
$$\beta_0, skip(X'_1), \beta_1, \ldots, skip(X'_m), \beta_m$$

where $nt \in V_N$; the α_i and $\beta_i \in V^*$; the skip(X_i) and skip(X'_i) $\in K$; $n, m \geq 0$. Prolog calls can be included too, but we shall disregard them for the time being.

For instance, the DG rule:

(1)   a, skip(X), b, skip(Y), c --> skip(Y), c, b, skip(X).

can be applied successfully to either of the following strings:

$$a, e, f, b, d, c$$

with skips X=e,f and Y= d, and

$$a, b, d, e, f, c$$

with skips X= [] and Y= d,e,f.

Application of the rule yields, respectively

---

[1] Originally, discontinuous grammars were called *gapping grammars*, but they were renamed because the term *gap* is used in linguistics with a different meaning than ours.

[2] XG's require that the two gaps be either totally independent or that one of them be completely included in the other.

dcbef

and

defcb

We can therefore think of the above DG rule as a shorthand for, among others, the two rules:

(1a)   a,e,f,b,d,c --> d,c,b,e,f

(1b)   a,b,d,e,f,c --> d,e,f,c,b

If we instead have the rule

a, skip(X), b, skip(Y), c --> skip(X), c, b, skip(X).

the first string becomes

efcbef

Notice that discontinuous grammar rules can be viewed as shorthand for several ordinary (logic grammar) rules, as the above example shows. In this sense they can be considered as metarules. But they are metarules in that their precise instances are constructed dynamically, as they are needed during the parse of a given sentence. Logic grammar rules could already be considered as metarules, since they stand for all instances obtained by substituting terms for the variables that appear in the arguments of grammar symbols. DGs generalize this idea, by also allowing variables (or, as we shall see later, unnamed strings) to stand for finite strings of grammar symbols.

Thus, DGs provide a great degree of economy, in two senses: not only do we not routinely produce specific rules like (1a) and (1b) that might never be used (and that may be infinite in number!), but the use of skips allows us a high degree of generality by which we can collapse what would be several rules in ordinary grammars into one.

Moreover, our metarules follow the same notation as the ordinary rules, with only the addition of the "skip" symbol, whereas in other approaches, metarules have their own, ad-hoc representation. Thus, we can consider ordinary rules as special cases of the general discontinuous rule format, in which there are no skips. In fact, DGs subsume most of the existing logic grammar formalisms as special cases.

It should also be noticed that there seems to be some psychological reality to the idea of skipping intermediate strings. Often, while analyzing a sentence, we focus on the next relevant substring, leaving an intermediate one suspended in the background of consciousness, to be brought back into focus later, possibly repositioned with other more closely related substrings. Appendix II shows a compiler for discontinuous grammars.

2. Thinking in Terms of Skips — Some Examples

A skip can be imagined as an unknown string of constituents to be repositioned (or copied, or deleted) by application of the discontinuous rule. We next examine several natural as well as artificial language processing problems in terms of discontinuous grammar rules.

2.1. Coordination

The following sample grammar handles sentence coordination and reconstitutes the meaning representation of an elided object.

    sentence(and(S1,S2)) -->sent(S1), and, sent(S2).
    sent(S) --> name(K), verb(K,P,S), object(P).

    object(P) --> determiner(X,P1,P), noun(X,P1).
    object(P), and, skip(G), object(P) --> [and], skip(G), object(P).

    determiner(X,P,the(X,P)) --> [the].

    noun(X,train(X)) --> [train].

    name(mary) --> [mary].
    name(john) --> [john].

    verb(X,Y,saw(X,Y)) --> [saw].
    verb(X,Y,heard(X,Y)) --> [heard].

As in metamorphosis grammars, a normalizing rule of the form

$$nt-->[nt].$$

is necessary for each nonterminal symbol *nt* contained in the left hand side of a rule. As these rules can be constructed automatically by a grammar preprocessor, we shall consider them transparent to the grammar writer and disregard them.

The second rule for "object" elides an expected object followed by a skip and reconstructs its internal representation *P* through unification with the meaning representation of the object in the second sentence. The skip in between "and" and the object is recopied unanalyzed.

A derivation graph for the sentence "mary saw and john heard the train" might help visualize the working of the grammar, shown in figure 1. The string "name(K2) vb(K2,P2,S2)" is the skipped substring G.

The final value obtained is

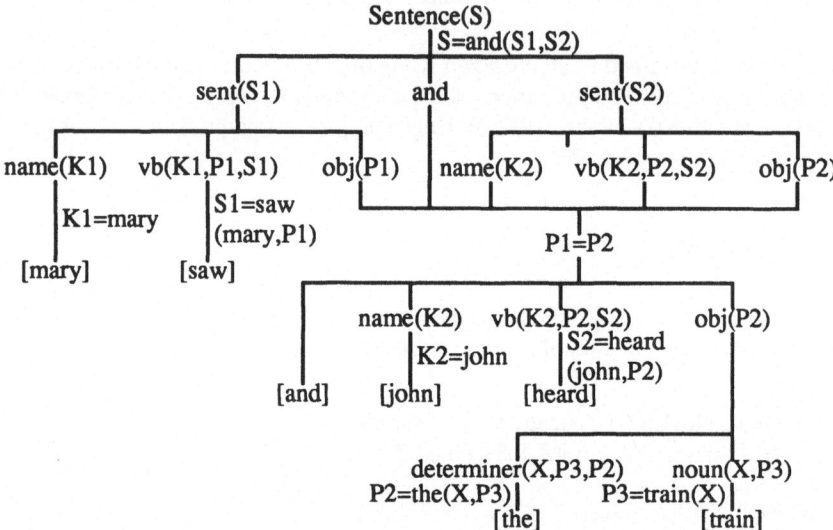

Figure1.
   Example of Coordination.

   S=and(saw(mary,the(X,train(X))),heard(john,the(X,train(X))))

2.2. Right Extraposition

We already saw, in section 3 of chapter 6, how extraposition grammars can be used to naturally describe left-extraposition phenomena, as in relativization. But in natural language, some movement phenomena are more naturally viewed as right rather than left extraposition, although they could perhaps be forced into left-extraposing formulations.

Before considering right extraposition, we shall recall, in our new notation with skips and in a version which includes all arguments, how to express relativization. (In chapter 6 only a skeleton derivation was shown.)

Relative clauses often can be viewed as mere canonical sentences whose parts have undergone changes and movement. Thus, a sentence such as "The man that

Jill saw is here'' can be viewed as the result of transforming

The man [Jill saw the man] is here

by moving the second occurrence of ''the man'' to the beginning of the embedded clause and replacing it with a pronoun. In order to parse sentences such as the above sentence into logical structures such as

the(X,and(man(X),saw(jill,X)),here(X))

we can define the following grammar:

sentence(P) --> np(X,P1,P), vp(X,P1).

np(X,P1,P) --> det(X,P2,P1,P), noun(X,P3), relative(X,P3,P2).
np(X,P,P) --> name(X).

vp(X,P) --> trans-verb(X,Y,P1), object(Y,P1,P).
vp(X,P) --> aux(be), comp(X,P1,P).

relative(X,P1,and(P1,P2)) --> rel-marker(X), sentence(P2).
relative(X,P,P) --> [].

rel-marker(X), skip(G), trace(X,P1,P) --> rel-pronoun, skip(G).

object(X,P,Q) --> np(X,P,Q).
object(X,P,P) --> trace(X).

comp(X,P,P) --> adverb(X,P).

noun(X,man(X)) --> [man].

aux(be) --> [is].

adverb(X,here(X)) --> [here].

det(X,P1,P2,the(X,P1,P2)) --> [the].

rel-pronoun --> [that].

name(jill) --> [jill].

trans-verb(X,Y,saw(X,Y)) --> [saw].

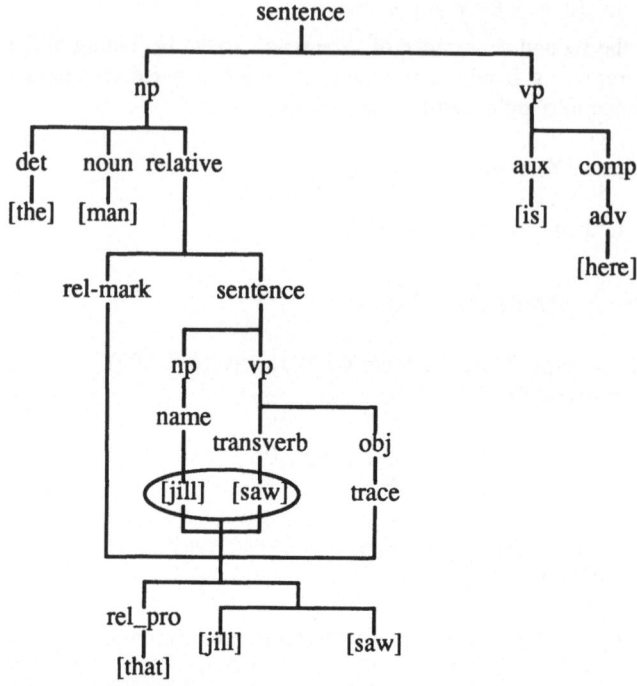

Figure 2. Relativization

The technique used is to rewrite a relative clause into a marker followed by a (canonical) sentence. The marker is then used, together with a trace left for the elided object, to perform the contextual changes mentioned. Graphically, we can view the rewriting as in figure 2, leaving arguments out for clarity. The string inside the ellipse stands for the skip in the *rel–marker* rule.

We next augment the little grammar shown above by adding two rules for extraposing the whole relative clause to the right. Both the left and the right extraposing rules can interact in the same sentence, as the example shows. The added rules are:

relative(X,P1,P), skip(G) --> skip(G), rightex(X,P1,P).

rightex(X,P1,and(P1,P2)) --> rel-marker(X), sentence(P2).

(The first of these rules, as we shall see in section 4, is needed to avoid loops.) The grammar now parses the sentence "The man is here that Jill saw" into the same representation as its paraphrase.

Exercises

1. Draw a derivation graph for "The man is here that Jill saw" □

2.3. Interactions between Different Discontinuous Rules

Consider the language $L(G) = \{a^n b^m c^n d^m : n, m \in N\}$, which can be described by the following DG:

s --> as, bs, cs, ds.

as -->[].
as, skip(X), xc --> [a], as, skip(X).

bs --> [].
bs, skip(X), xd --> [b], bs, skip(X).

cs --> [].
cs --> xc, [c], cs.

ds --> [].
ds --> xd, [d], ds.

This is a perfectly good DG. XGs cannot, however, be used in this situation because of the XG constraint on the nesting of skips: two skips must either be independent, or one skip must lie entirely within the other.

2.4. Avoiding Artifices through Straightforward Uses of Skips

The grammar above uses marker symbols *xc* and *xd*, whose only function is to leave traces of right-extraposed a's and b's in place where they can easily be evened out with c's and d's, respectively. However, thinking in terms of markers is in many cases simply a residue from the constraints imposed by less powerful grammars.

Thus, the language $L(G) = \{a^n b^m c^n d^m : n, m \in N\}$, for which a formulation with markers was shown above, can be more easily formulated as follows:

    s --> as, bs, cs, ds.

    as, skip(G), cs --> [a], as, skip(G), [c], cs.
    as, skip(G), cs --> skip(G).

    bs, skip(G), ds --> [b], bs, skip(G), [d], ds.
    bs, skip(G), ds --> skip(G).

The second rule simply evens *as* and *cs* with *a* and *c*, by skipping any intermediate string as a skip G, which is repositioned after regenerating *as* and *cs* in the right hand side. The third rule simply makes *as* and *cs* vanish.

2.5. Rules with More Than One Skip

Using the same technique as for the above grammar, we can describe the language $L(G) = \{a^n b^n c^n\}$ through:

  1) s --> as, bs, cs.

  2) as, skip(G1), bs, skip(G2), cs -->
     [a], as, skip(G1), [b], bs, skip(G2), [c], cs.

  3) as, skip(G1), bs, skip(G2), cs --> skip(G1), skip(G2).

2.6. DGs and Free Word Order Languages

Many languages exhibit, to some extent, case marking through morphology rather than through position in the sentence. Since, in these languages, altering the position does not make the case dubious, some constituents can freely move from their "orthodox" positions into arbitrary or fairly arbitrary locations, as a matter of style, emphasis, etc...

2.6.1. Totally Free Word or Constituent Order

For instance, the Sanskrit phrase "Ramauh pashyati Seetam" (Ramauh sees Seetam) can also appear as

>   pashyati Ramauh Seetam
>   pashyati Seetam Ramauh
>   Seetam Ramauh pashyati
>   Seetam pashyati Ramauh
>   Ramauh Seetam pashyati

This kind of free order of sister constituents where each retains its integrity is easily handled within discontinuous grammars, in a similar way as our last example (e.g., sample rule in section 2.5). More interesting is the case in which even the contents of constituents appear to be scrambled up with elements from other constituents (e.g., as in the Walpiri language). Even in Latin or Greek, phenomena such as discontinuous noun phrases, which would appear as extreme dislocation in prose, are very common in verse (and not unusual even in certain prose genres, e.g. Plato's late work, such as the *Laws*). A contrived example for Latin would be:

>   Puella bona puerum parvum amat. (Good girl loves small boy)

where the noun and adjective in the subject and or object noun phrase may be discontinued, e.g.

>   Puella puerum amat bona parvum.

In fact all 5! word permutations are possible, and we certainly do not want to write a separate rule for each possible ordering. In DGs, we can simply write

sentence --> noun-phrase(nom), noun-phrase(acc), verb.

noun-phrase(Case) --> adjective(Case), noun(Case).

noun(Case), skip(G) --> skip(G), [Word], {dict(noun(Case),Word)}.

adjective(Case), skip(G) --> skip(G), [Word],
                    {dict(adjective(Case),Word)}.

verb, skip(G) --> skip(G), [Word], {dict(verb,Word)}.

dict(verb, amat).
dict(noun(acc),puerum).
dict(noun(nom),puella).
dict(adjective(acc),parvum).
dict(adjective(nom),bona).

Notice that the number of rules needed grows linearly depending on the number of constituents that can freely move.

Another approach is the augmented phrase structure (Pullum 1982). In Pullum's formulation, phrase structure (meta) rules only indicate immediate dominance, and are supplemented with linear precedence restrictions to indicate what orderings are allowed. For instance, the metarule

A --> B, C, D.

together with an empty set of linear precedence restrictions, stands for all rules where A rewrites into B, C, and D in any order. With the restriction: {D<C}, on the other hand, it represents only the rules:

{A--> BDC, A -->DBC, A--> DCB}.

While this notation is concise and expressive for free or relatively free ordering problems, it becomes costly as more orderings are fixed. Also, since precedence restrictions are attached to the whole set of phrase structure rules, it can only deal with grammars in which any two constituents that have been stated to have an order appear in that order no matter what their origin. Gazdar and Pullum make the hypothesis that grammars of natural language will all possess this property.

DGs, on the other hand, can describe different orders of the same constituents coming from different rules quite straightforwardly, so they will be appropriate even if the above mentioned hypothesis turns out to be wrong, or if we want to process formal language grammars that do not satisfy it. They also seem more versatile in being able to deal with both fixed and changing order with no significant change in cost.

2.6.2. Lexically Induced Rule Format and Free Word Order

Another interesting problem that can be handled in DGs is free order of consti-tuents that may or may not be present, as determined by other constituents. Let us assume that each particular verb requires its own set of constituents. Then we could include an argument in the verb symbol, telling us about its specific requirements. Modulo notation[3] , this would look like:

sentence --> verb(R), R.

verb(R), skip(G) --> skip(G), [W], {ver(W,R)}.

nominative, skip(G) --> skip(G), [W], {nom(W)}.

accusative, skip(G) --> skip(G), [W], {acc(W)}.

dative, skip(G) --> skip(G), [W], {dat(W)}.

ver(pashyati,[nominative,accusative]).

ver(yachchhati,[nominative,accusative,dative]).

nom(ramahuh).
acc(seetayai).
dat(pushpam).
nom(ramauh).

acc(seetam).

This grammar accepts

> Ramauh pashyati seetam.
> Ramahuh Seetayai pushpam yachchhati.
> (Ramahuh gives a flower to Seeta)

The *sentence* rule first generates the fixed order *verb, . . . ,* where what follows the verb will be known as a result of processing the *verb* symbol. This, in turn, is done through skipping any preceding constituents until the verb (W) is found. Its lexical definition has a subcategorization frame (R) associated to it. By virtue of the *sentence* rule, this frame materializes as part of the derivation tree.

The possibility that a rule takes a format specifically suited to the requirements of given words might be useful in theories such as Government and Binding, in which the argument structures required by certain words are described in the lexicon. The frame requirements may be associated with a word appearing in any position, even preceding the actual materialization of one of the requirements. For instance, a nominative constituent required by a verb may well appear before that verb, but discontinuous rules may allow us to process the verb beforehand.

---

[3] Normally the variable R above will unify with a list of constituents, so it cannot be directly used as a metacall in the first rule. Further processing would be needed, but we represent it here as a metacall for simplicity.

F. Popowich (1985) presents a method to deal with partially free word order within discontinuous grammars. In particular, he shows how to express the immediate dominance/linear precedence format rules (Gazdar et al. 1982) by a set of unrestricted (i.e., with the starting symbol being either a nonterminal or a terminal) discontinuous grammar rules that contain procedural control.

3. Static Discontinuity Grammars and Government-Binding Theory

This section will be especially useful for readers who come from the linguistics field. If other readers find it too specialized, it may be skipped with no loss in continuity.

V. Dahl's efforts on merging Chomsky's Government–Binding (GB) Theory with the Logic Grammar paradigm started in 1984, from the observation that, in general aims and philosophy, discontinuous grammars present some exploitable similarities to GB theory.

Both attempt to achieve maximal coverage with minimal expressive machinery, both can describe fairly free types of movement, both tend toward abstraction and generalization. Together with the similarities, of course, there are admitted disadvantages in trying to use GB theory for natural language processing: GB is not fully formalized, it stresses the generative paradigm, it is an evolving rather than a solidified theory. But many of its aspects can be usefully adapted into computational linguistics applications, and logic programming is proving to be relevant in this respect.

One of the obvious features that make logic grammars appropriate to GB theory is the ease with which abstraction on head constituents can be performed: GB theory generalizes head categories such as "noun," "verb," "adjective," into an abstract category called X, by means of which it collapses all rules describing phrases with a specific head (i.e., all noun phrases, adjective phrases, verb phrases, etc.) into basically just two rules describing X-phrases in general. This is very naturally expressed in logic grammars by making the category ("noun," "adjective," "verb," etc.) be an argument of that generic "head" symbol X.

3.1. Rendering Context-Free Simplicity with Type-0 Power

Another exploitable feature of logic grammars, this time specifically of discontinuous grammars, is the ease with which movement can be described very generally, by highlighting only those constituents that move and making the intermediate ones be represented by skips.

However, for GB theory as well as for other linguistic theories, Discontinuous grammars had the disadvantage of inducing parse graphs rather than trees. Hierarchical relationships were lost in a rule's flattening of multilevel symbols into siblings, thus making it difficult to state the typically hierarchical principles of Government–Binding theory.

In static discontinuity grammars, tree structure is preserved while not giving up type-0 power. This is achieved by the static discontinuity feature, which forbids the skips to move and allows only the explicit constituents around them to do so. Thus, we can write a rule schema such as:

empty,skip,y --> y,skip,trace(y).

to achieve movement of a category *y* to an empty position (as in move-alpha).[4] Notice that the movement is described statically, by rewriting *empty* into *y* and *y* into *trace(y)*. Since the skipped substring does not move, two branches can depict the rule's expansion, each with a single parent:

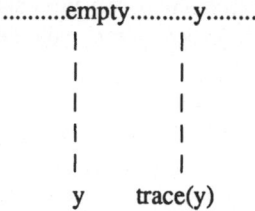

The skipped substrings need not in fact be even noted in the rule schema, which can alternatively be expressed as the conjunction of the two subrules:

empty --> y,
y --> trace(y)

affected by the same substitutions. In section 3.3 we show a formal definition of the general SDG formalism. Type-0 power is achieved by the fact that the substitutions used must be the same for all subrules, so that although they look like a set of context-free rules, the shared substitutions, as well as the context provided for each subrule by all other subrules, ensure context-sensitive and transformational power. Movement can be described statically, in terms of rewriting from a single parent, while all relevant elements remain related through substitution.

Static discontinuity grammars are at the origin of two Government-Binding systems developed at Simon Fraser University: a machine error message generator and a grammar of Spanish with clitic treatment.

---

[4] The move-α rule allows arbitrary movement into any empty position. Ungrammatical sentences overgenerated by move-α are ruled out by filters applied at a later stage.

3.2. Government–Binding – Oriented Constraints

Constraining mechanisms for discontinuous grammars in general were described in work by Dahl and Saint-Dizier. Basically, this research augments the compiler shown in appendix II with constraints expressed as the interdiction to move any node outside a specified domain. But, although it did allow to express some linguistic constraints and thus reduce the generation of ungrammatical sentences, this first attempt had several drawbacks, namely: a) it worked only for analysis, and not for generation; b) parse graphs, rather than trees, were produced, thus requiring an inelegant artifice to keep track of hierarchical relationships, often crucial to linguistic constraints in general; c) a further artifice was needed to avoid loops; d) skipped substrings were tried in the inefficient order of increasing length, causing in general much backtracking, and e) the constraint format was not specifically tailored for the typical constraints of Government–Binding theory.

These problems were solved by V. Dahl's introduction of the more easily implementable SDG formalism. In particular, by allowing movement (or relating) of discontinuous constituents to be statically described, and by therefore elliciting parse trees rather than graphs, SDG constraints can represent any linguistic filter that can be expressed in terms of node domination within a (current) parse tree status. Efficient primitives for checking node dominantion in constant time have been described by Dahl and Massicotte (1988). For our purposes here, we shall only present the GB-oriented version of our constraining mechanism, which has the following format:

constraint(Path,Node,Root) :- body

where **Path** describes a path in a derivation, **Node** a node in that path under which no element can move out of a given zone, and **Root** the root of that zone. **Body** is a list of Prolog calls, which can of course be empty.

We shall now see how to implement the GB principle of subjacency using these constraints. Subjacency identifies a language-dependent concept, that of **bounding nodes,** which serves to block forbidden movements in a given language. For instance the bounding nodes for English are **sentence** and **noun-phrase**, the bounding nodes for Italian are **sentence-bar (complementizer phrase)** and **noun-phrase**, etc.[5] Subjacency simply states that the movement rule (move-α) cannot cross more than one bounding node at a time. The movement rule, in turn, allows moving into specific, empty positions. Thus, the direct object in the sentence *John likes ___* embedded in the following sentence:

Who do you believe John likes?

has been allowed to move to the head of the sentence in the form of the pronoun *who*, by means of first crossing the *sentence* node of its immediate subsentence,

---

[5] These constituents are respectively called "s," "n double bar," and "s bar" in Government-Binding theory; we use less technical names here to suit a wider audience.

and then crossing the main *sentence* boundary, i.e.,

[$_{S-bar}$ Who [$_S$ do you believe [$_{S-bar}$ ___ [$_S$ John likes ___]]]]

In contrast, the ungrammatical sentence *Who do you wonder why John likes?* is ruled out by subjacency, since the direct object would have to move across the two *sentence* nodes at once: the position under *s-bar* that it moved to in the previous sentence is already taken by *why*. In graphic terms:

*[$_{S-bar}$ Who [$_S$ do you wonder [$_{S-bar}$ why [$_S$ John likes ___]]]]

Subjacency can in general be viewed as the interdiction for a rule to move any constituent **X** to a position outside a domain **a** in the situation

$$[_a \ldots [_b \ldots X \ldots] \ldots]$$

where both **a** and **b** are bounding nodes.

An intelligent parser that could look at its derivation tree could enforce such a constraint by allowing users to represent the configuration shown above, and indicate that no descendant of **b** can move outside **a** in such a configuration.

For our system, this is written:

constraint([A...[B...]...],B,A) :- bounding(A),bounding(B).

which is read "if a bounding node **A** has as descendant a bounding node **B**, no descendant of **B** can move outside the domain with root **A** ."

Subjacency is thus expressed in a single, very general rule. In order to adapt it to different languages, we simply add, for instance

bounding(s).
bounding(np).

if the language is English, or

bounding(s-bar).
bounding(np).

if we are dealing with French, Italian, etc.

With respect to a given grammar, constraints act as inhibitors of any rule whenever it attempts to perform a movement forbidden by the constraint. Naturally, the associated checking and eventual inhibition take place automatically, as does the bookkeeping activity of recording the derivation so far.

3.3. Definition

A *static discontinuity grammar* (SDG) is a logic grammar $G=(Vn,Vt,P,s)$ where Vn,Vt and s have their usual meanings, and P is a set of productions, each of one of the two following forms:

Type-1 productions:

$nt \rightarrow \beta$.

Type-2 productions:

$nt_1 \rightarrow \beta_1,$
$nt_2 \rightarrow \beta_2,$

$\cdot$
$\cdot$
$\cdot$

$nt_n \rightarrow \beta_n.$

where nt_i are nonterminals, and β_i are sequences of terminals and nonterminals (Prolog calls may be included too, but we shall disregard them here.) Rules of the second form (discontinuous rules) as can be seen, consist of various context-free like rules. They can be used to rewrite strings of the form

$nt'_1 \, s_1 \, nt'_2 \, s_2 \, ... \, nt'_n$

into

$$(\beta_1 \, s_1 \, \beta_2 \, s_2 \, \cdots \, \beta_n) \, \tau$$

where nt_i unifies with nt'_i with a most general unifier τ, and s_i are strings of terminals and nonterminals, and no constraint prevents the production's application. Constraints are defined as in section 3.2.

Because the strings s_i are not moved, the corresponding part of the derivation can be depicted by drawing an arc from nt'_i to β_i for every i. Graphically:[6]

```
.... nt'₁  s₁  nt'₂  s₂  nt'ₙ .......
      |         |          |
      |         |          |
     β₁τ       β₂τ        βₙτ
```

Thus, although the parsing depiction remains a tree, context-sensitive and transformational power are explicitly present in the sharing of the substitution τ by all symbols concerned, and in the fact that subrules provide context to each other.

3.4. Some Implementation Considerations

Several implementations have been used in our applications of GB theory. The SDG formalism itself does not presuppose any particular strategy for applying the subrules of an SD rule, although specific applications will adopt one. Subrules can be seen as a set or a list, they can be applied in parallel (i.e., only when a complete matching string is found in a parsing state) or in successive steps that keep track of the substitutions to propagate and of the unapplied subrules, etc. Different strategies determine specific variants of SDGs. Section 2 of appendix II presents an interpreter that uses the latter of the above strategies, while providing a parallel flavor whenever possible (Dahl and Massicotte 1988). The reader is encouraged to test it and to modify it in order to try different subclasses of SDGs. An efficient scheme for implementing constraints is described in the same article.

3.5. Discontinuous Logic – Transporting the Static Discontinuity Feature into Logic Programming

The static discontinuity paradigm is not limited to logic grammars: we can also apply it in logic programming proper, thus giving rise to Discontinuous logic, in which strings of terms in a proof state can be skipped while expanding other

---

[6] For explanatory purposes, it is convenient to visualize the application of all subrules together, but the reader should bear in mind that – just as in the general DG framework – other strategies are possible and desirable, as is discussed in the next section.

explicit, discontinuous terms in that state, with shared substitutions. Discontinuous logic rules can thus be thought of as logic programming productions that are grouped to provide context to each other and to share substitutions. Such productions can be noted, for instance:

$$P_1 :- \beta_1,$$

.

.

.

$$P_n :- \beta_n.$$

The P_i are (discontinuous) clause heads and the β_i are the corresponding bodies. The application of such a production to a proof state $\cdots P'_1 s_1 P'_2 s_2 \cdots P'_n \cdots$ where the P_i match the P'_i with substitution θ, results in the new proof state

$$(\cdots \beta_1 s_1 \beta_2 s_2 \cdots \beta_n \cdots)\theta.$$

The interpreter shown in appendix II, section 2 can be easily adapted to discontinuous logic programming yielding an even shorter program–one clause less. This is because terminal symbol recognition is no longer necessary.

Of course, discontinuous logic programming can also incorporate constraints like those examined for SDGs, with similar implementation considerations. It should also be noted that, here again, different strategies yield different variants of the formalism, with different practical implications, and that subrules may not all apply on the same parsing state, although this is a convenient way of visualizing them.

Bibliographic Commentary for Part III

The semantic interpretation component of modifier structure grammars is, except in the aspects involving coordination, mostly taken from McCord (1981, 1982) where its description can be found. The aspects involving coordinate modifiers as well as all the other points covered in chapter 8 are described in Dahl and McCord (1983).

The first processor for discontinuous grammars was a compiler devised by Dahl and later coded by McCord during their joint work on coordination (cf. appendix II).

Subsequent work on implementation of discontinuous grammars was done jointly by the authors (Dahl and Abramson 1984) and by Popowich (1985); its uses and further possibilities were investigated in Dahl (1984) and in Popowich (1985).

The static discontinuity feature, and its associated Government-Binding oriented constraint mechanism, were introduced by Dahl (1986), motivated by her ongoing research on using that linguistic theory as a formal framework for processing natural language. Its uses are, however, neither restricted to language applications nor to GB theory alone. Further details can be found in Dahl (1988a, 1988b).

Other approaches to capture Government-Binding theory within logic programming are that of Sharp, which uses strictly logic programming (Sharp 1985), that of Stabler, based on restricting logic grammars through constraints on local derivations in the parser (Stabler 1987, 1988), and that of Johnson, which applies program transformations to the construction of parsers obeying some of the contraints of GB theory (Johnson 1987, 1988). Scattered context grammars (Greibach and Hopcroft 1969) can be considered a non-logical grammar antecedent of SDGs, in which grammar symbols have no arguments and subrules must apply simultaneously. They describe a subset of context-sensitive languages, whereas in SDGs full context-sensitive and transformational power is preserved.

SDGs have originated two Government–Binding systems developed in Dahl's research group: a machine error message generator and a grammar of Spanish with clitic treatment. Material related to our adaptation of Government–Binding theory is covered in Brown, Dahl et al. (1986), and other material relevant to these applications, in Brown, Pattabhiraman et al. (1986); Brown (1987); Brown et al. (1988); and Massicotte et al. (1988). A subclass of Static Discontinuity grammars, interpreting subrules specifically as sets and adding linear precedence restrictions and modalities, was also studied at Simon Fraser University (Saint-Dizier 1988). The error message generator uses (our own adaptation of) conceptual graphs (Sowa 1984). A discussion of logic grammar and linguistic theories can be found in Dahl et al. (1988).

In early reports, the SDG formalism was either unnamed (Dahl 1986), or was called (inappropriately) Constrained Discontinuous grammars (Dahl et al. 1986; Brown 1987). To avoid confusion, the misnaming in these reports has been corrected by the authors once a suitable name was coined, since Constrained

Discontinuous grammars is in fact a different, existing formalism (Dahl and Saint-Dizier 1986), which simply augments general Discontinuous grammars with constraints, and does not possess the static discontinuity feature. Its constraints, as described in section 3.2, have served as ancestors of SDG constraints, but are in fact quite different, since, among other things, the latter exploit the feature of unmovable skips for achieving power that cannot be attained in Constrained Discontinuous grammars.

The idea of discontinuous logic, which adapts the static discontinuity paradigm to the logic programming framework rather than the grammatical one was introduced in Dahl et al. (1986) and followed up in Saint-Dizier (1987), where it is found renamed as DISLOG. Implementation issues for the discontinuous framework in general and the static discontinuity framework in particular are discussed in Dahl and Massicote (1988).

Since the introductory publication of Knuth (1968), a vast literature has developed on Attribute Grammars. There is a three-part survey of the literature in Deransart, Jourdan, Lorho (1985, 1986, 1986b) covering the main results on Attribute grammars, and providing a comprehensive review of existing systems and a classified bibliography.

The section on compiling was of course inspired by the pioneering work of Colmerauer (1975) on Metamorphosis Grammars which contained a grammar for a small programming language, and Warren (1980) which showed how compiling can be handled within Definite Clause Grammars. Our approach, using Definite Clause Translation Grammars, is exactly that of syntax directed translation where syntax and semantics are separated for modularity and ease of understanding. Syntax directed translation is used as a declarative method of compiler specification in most textbooks (see, for example, Aho and Ullman [1972, 1977] and Aho, Sethi, Ullman [1986]), but procedural specification usually degenerates into the use of a traditional programming language. It is our belief that all aspects of compiling technology can profitably be expressed logically. Some examples of this appear in Cohen and Hickey (1987), which in addition to covering an example similar to our compiled language (but, somewhat regressively, with DCGs where syntax and semantics are not strictly separated), also Prologically expresses a number of algorithms related to discovering properties of context-free grammars. Also related to the topic of expressing compiler technology in logic is Nilsson (1986), where it is shown how to implement DCGs when the underlying context-free grammar is SLR(1). Valuable sources of information on the use of logic to define programming languages may be found in Moss (1980, 1981, 1982). A purely functional programming language HASL and its interpreter are specified logically in Abramson (1986). The specification is divided into phases for lexical analysis, syntactic analysis, code generation and optimization, and interpretation of the generated code. There is a description of a logically based translator writing system in Abramson et al. (1988), where the user has available an array of parsing methods for grammars specified by DCTGs and a simple expert system which, on the basis of properties of the underlying context-free grammar of the language being implemented, chooses the appropriate parsing method; in addition, the system contains optimization and code generation algorithms, and the expert system is to be gradually extended so that the user can at least have mechanical aid in putting together a translator for a programming language.

Logic programming contains the possibility of using other procedural interpretations than the sequential one of solving goals in a strict order. Several Prolog implementations permit a kind of coroutining between the solution of goals, depending on whether some variable or variables have been bound. Using such an implementation of Prolog would permit, for example, the concurrent execution of lexical analysis, syntactic analysis, and translation (attribute evaluation). One such possibility is mentioned in Cohen and Hickey (1987) in connection with use of the *freeze* predicate where evaluation of *Predicate* is suspended in case *Variable* has not been instantiated:

    freeze(Variable,Predicate).

However, *freeze* is perhaps too low-level a method of obtaining coroutining in that the user has to explicitly freeze goals to be suspended. A far more interesting avenue of exploration lies in the use of automatically generated control information based on the work of Naish (1985) and implemented in NU-Prolog.

There are interesting connections between logic programs and the two-level grammars (mentioned in the section on metarules) which are explored in Maluszynski (1982) and Maluszynski and Nilsson (1982b). The grammatical data typing described here was inspired in part by Maluszynski and Nilsson (1981, 1982a). Two-level grammars and their applications are discussed in Cleaveland and Uzgalis (1977).

Proposals for other parsing strategies for DCGs are discussed in Stabler (1983), Porto et al. (1984), Matsumoto et al. (1983), and Uehara et al. (1984); bottom-up parsing is discussed in more detail in chapter 11. A logic programming language called ESP has been proposed for for object-oriented parsing (Miyoshi et al. 1984).

PART IV

OTHER APPLICATIONS

Chapter 11

Other Formalisms

1. Restriction Grammars

Restriction Grammar is described in Hirschman (1986) as a "grammar-writing framework in Prolog." The Restriction Grammar framework (Sager, 1967, 1981; Sager and Grishman 1975), following its origins in Sager's String Grammars, consists of BNF (context-free) definitions for syntax, coupled with constraints or *restrictions* on the structure of the parse tree. The RG formalism automatically constructs the parse tree: thus it differs from the Metamorphosis Grammar and Definite Clause Grammar formalisms which do not, but not from later logic grammar formalisms such as Modifier Structure Grammar and Definite Clause Translation Grammars which do. The portion of the Restriction Grammar that is not context-free is specified by "restrictions" on the structure of the parse tree, rather than by augmentation of the nonterminal symbols with extra arguments as in Definite Clause Grammars.

Since the syntactic portion of Restriction Grammar rules consists of BNF definitions, nothing further need be said about it. The restrictions however require further comment. A typical Restriction Grammar rule looks like:

$$predicate ::= verb,object,\{verb\text{-}object\}.$$

This rule states that *predicate* may be rewritten into *verb* followed by *object* provided that the *verb−object* restriction is satisfied. The restriction might state, for example, that if the object is empty, then the verb must be intransitive. The *verb−object* restriction has no explicit parameters, but there are the implicit parameters of the current point in the tree and the current list of input words. These parameters are hidden from the Restriction Grammar writer by the system. Restrictions are applied as soon as the node immediately to the left has been constructed. In the above rule, as soon as *object* has been recognized, the *verb−object* restriction is applied.

Syntactically motivated restriction routines, such as *head, adjunct,* etc., are written in terms of primitive tree relations which support restriction language operators. Such routines in conjunction with the primitive level operators are used to implement general restrictions. A data structure such as:

link(TreeTerm, Path)

supports the most primitive level of tree relations. Here, **TreeTerm** stands for the current node of the parse tree, and **Path** represents the way to get back from the current node to the root of the tree. **TreeTerm** is a data structure of the following form:

tt(Label, Child, RightSib, Word)

where **Label** is the node name given by a nonterminal in a BNF definition, **Child** and **RightSib** are themselves **TreeTerms** representing the first (leftmost) child and right sibling respectively of the current node, and **Word** contains the lexical item and its attributes. The **TreeTerm** structure is recursive and makes it possible to access daughter nodes (or son nodes, so as not to be sexist!) and also right siblings of the current node.

A **Path** to the root of the parse tree is a data structure consisting of a sequence of the binary functors **up** and **left** specified as follows:

link(TreeTerm, up(Parent, ParentPath))
link(TreeTerm, left(LeftSib, LeftSibPath))

The second argument to **up** and **left** is the remainder of the path to the root.

The basic tree relations are **down, right, up** and **left.** To give a flavor of how these are specified, here is the **down** relation which finds the daughter relation:

```
down(link(TreeTerm,Path),link(NewTreeTerm,up(TreeTerm,Path))) :-
    TreeTerm = tt(_,NewTreeTerm,_,_),nonvar(NewTreeTerm).
```

The first argument represents the current node and is assumed to be instantiated, while the second argument is instantiated to the new node. The new path from the daughter is returned as **up(TreeTerm, Path).** (See Hirschman [1986] for a complete specification of these relations.) These relations have the highly procedural flavor of the original (Fortran!) Linguistic String Project implementation (Sager and Grishman 1975), but it is expected that as the current implementation proceeds, these relations will take on a more declarative aspect.

Besides these primitives which permit tree traversal, there is the **label(Node, Name)** relation between a node and the nonterminal naming it, and also the **word(Node, Word)** relation between a node and the word or words it encompasses.

Restriction operators are one layer above the primitive relations, and permit, for example, scanning of the current input word or the next input word, looking ahead through the input string for a particular word, checking to see whether a node is empty or not, and ascension or descension through nodes of the tree under certain specified conditions. From the primitive relations and these operators, restriction routines are implemented.

Restriction grammars are not, strictly speaking, logic grammars in the sense that the writer of a restriction grammar is not directly making use of the underlying resolution mechanism of Horn clause logic, nor is the writer able to directly make use of unification or the logical variable. Restriction grammars have their use, however, "because a very comprehensive English grammar exists in this framework" (Hirschman 1986).

Exercises

1. The Definite Clause Grammar formalism permits one to specify the syntactic portion of a grammar by context-free like rules, and the non–context-free portion by a set of Horn clause like rules for expressing constraints, restrictions, translations, etc. Examine a portion of a restriction grammar and re-express it declaratively in terms of the DCTG formalism. □

2. Puzzle Grammars

In puzzle grammars (ZGs), a language is defined through key-trees rather than rewrite rules (Sabatier 1984).

Key-trees have the form

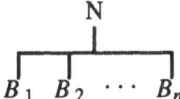

where N is a predicate, and each B_i is a terminal, nonterminal, the empty string, or another key-tree.

For instance, the language $\{a^n\ b^n\ c^n\}$ is described through the key-trees of figure 1.

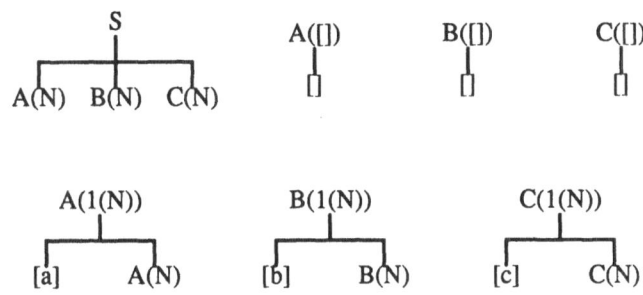

Figure 1.
        Puzzle Grammar for the language $\{a^n\ b^n\ c^n\}$

Constraints, in the form of Prolog calls, can be attached to a key-tree to restrict the way in which key-trees can be assembled. Assemblage of key-trees yield new key-trees. This is done by unifying roots and leaves of key-trees in order to "glue" them together.

In the above example, for instance, we can assemble two instances of the last tree with each other, yielding:

$$C(1(1(N1)))$$

```
        ┌──────┴──────┐
       [c]         C(1(N1))
                ┌──────┴──────┐
               [c]          C(N1)
```

This in turn we can assemble with the other tree with root C, whereupon N1 takes the value nil, and we have:

$$C(1(1([])))$$

```
        ┌──────┴──────┐
       [c]         C(1([]))
                ┌──────┴──────┐
               [c]          C([])
                              │
                              []
```

Similar trees can be obtained from the rules for A and B, and then we can assemble the three resulting trees with the first rule, giving:

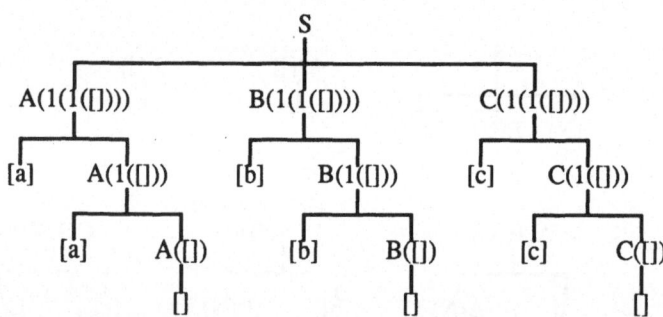

i.e., a parse tree for the sentence: $a^2 b^2 c^2$.

The idea is to express context dependencies by means of the tree structures they occur in, and to make the parsing strategy more flexible. In order to analyze/generate a given sentence from a given puzzle grammar, strategic rules select "interesting" key-trees for the assemblages, and govern the order and the mode (top-down, bottom-up, deterministic, nondeterministic, left-to-right, right-to-left, etc.) of assemblages.

Puzzle grammars are interesting in that they escape quite naturally the simple top-down, left-to-right Prolog strategy. They are also unique among logic grammars in that rewriting is no longer the leit-motif. However, the relative declarative transparency of rewriting rules seems hard to match in ZGs; trees are in fact written in functional notation, and it is not yet clear whether the control strategy will involve the grammar writer (thus introducing operational concerns into the grammar formalism) or be as general as to be kept invisible to the user (in which case, guaranteeing a reasonable efficiency in all cases might prove very difficult).

3. Discussion

Restriction grammars have been presented here as an implementation tool, within logic programming, for a well-known linguistic formalism. Linguistically oriented readers might, however, find this section of interest.

Puzzle grammars are more within the philosophy of logic grammars, but so far only interesting from a theoretical point of view, since a practical control strategy has not been properly determined.

Exercises

1. Examine the use of typing DCTGs for the specification and implementation of puzzle grammars. The emphasis in Typing DCTGs as in puzzle grammars is on tree structures rather than immediately on parsing or generation. □

Chapter 12

Bottom Up Parsing

1. Introduction

The declarative reading of a logic grammar specifies the strings that are in the language generated by the grammar; the procedural reading of a logic grammar specifies how strings of the language are generated or analyzed. Up until now we have used a top-down (with respect to the parse tree) procedural reading for our logic grammars: grammar rules are compiled to Prolog clauses which analyze or generate sentences starting from the root of the parse tree. We will now describe another procedural interpretation of DCTGs which in fact constructs the parse tree from the bottom up. This is based on the work of Matsumoto et al. which is called "BUP: a bottom-up parser in Prolog." One reason for considering bottom-up parsing is that top-down parsing prevents the use of left-recursive rules. The BUP parser allows the use of left-recursive rules, but is, however, unable to deal with rules with empty right hand sides. Another reason for considering bottom up parsing is that some scientists consider bottom up sentence recognition as more cognitively likely.

Consider the following simple grammar:

```
s ::= np, vp.
np ::= [pascal].
vp ::= [stinks].
```

Parsing of the sentence [pascal, stinks] proceeds in the following fashion. Initially, the goal is to parse an *s* given that the first symbol in hand is the atom *pascal*. A dictionary is consulted which tells us that pascal is an *np*, and other information tells us that it is possible to use an *np* in satisfying the goal of finding an *s*. Next, we establish *vp* as a new goal and proceed as before. Now, the symbol in hand, *stinks* is looked up in the dictionary and we find that it is a *vp*, and other information tells us that indeed it is possible to use a *vp* in satisfying the goal of finding a *vp*: indeed, this is the *vp* we are looking for! Having found a *vp* we have exhausted all the symbols in the putative sentence, but we have also done everything necessary to satisfy the goal of finding an *s* and we conclude that [*pascal,stinks*] is an *s*.

We may consider the Prolog clauses which implement the parsing method just described.

```
np(Goal,X,Z) :- goal(vp,X,Y),s(Goal,Y,Z).
dict(np,[pascal|S],S).
dict(vp,[stinks|S],S).
goal(G,X,Z) :-
  dict(C,X,Y),
```

```
    P =.. [C,G,Y,Z],
    call(P).
np(np,X,X).
vp(vp,X,X).
s(s,X,X).
```

The reader is asked to convince herself that given the Prolog goal:

```
?- goal(s,[pascal,stinks],[]).
```

the above Prolog program yields a positive solution to the query.

2. Compiling Context-Free Rules

We begin by showing how context free rules are transformed to Prolog clauses which specify the action of a left-corner parser, a variety of bottom-up analysis. These clauses will parse a sentence and automatically construct a parse tree.

Consider purely context-free rules of the following two forms:

rule 1:   nt ::= [terminal]

rule 2:   nt ::= $sym_1, \ldots, sym_n$ $(n \geq 1)$

where *nt* is a nonterminal symbol, *[terminal]* represents a terminal symbol, and where $sym_i (1 \leq i \leq n)$ are either terminal or nonterminal symbols.

A rule with only a single terminal symbol on the right hand side is treated as a dictionary rule. The unit clause generated for such a rule looks like:

```
dict(nt,node(nt,[[terminal]],'[]'),[terminal|X],X).
```

The first argument specifies the left-hand side of the grammar rule, i.e., what kind of nonterminal. The second argument specifies the subtree generated during parsing by use of this rule, and the last two arguments represent the difference list of the string being analyzed or generated, and that the first element of the string must be *terminal*. The predicate *dict* is essentially a more general version of the *connect* predicate used by the DCTG processor and represents a lookup of a word in a possibly large dictionary data base.

For the rules:

```
np ::= [pascal].
vp ::= [stinks].
```

the clauses generated are:

```
dict(np,node(np,[[pascal]],'[]'),[pascal|X],X).
dict(vp,node(vp,[[stinks]],'[]'),[stinks|X],X).
```

The other kind of grammar rule is more complex and gets translated into Prolog clauses of the following form:

```
sym1(sym1,Tree,Tree,X,X).              %rule 1

sym1(Goal,TreeOut,Tree1,X0,Xn) :-      %rule 2
    link_rtc(nt,Goal),
    goal(sym2,Tree2,X0,X1),
    goal(sym3,Tree3,X1,X2),
    ...
    goal(symn,Treen,Xn2,Xn1),
    nt(Goal,TreeOut,node(nt,[Tree1,...,Treen],[]),Xn1,Xn).
```

(Here $Xn2$ and $Xn1$ represent X_{n-2} and X_{n-1} respectively.) Grammar rules such as this one get translated into Prolog clauses with five arguments:

argument 1: the final goal to be attained during parsing.
argument 2: the portion of the dervivation tree constructed
            as a result of use of the Prolog rule.
argument 3: a subderivation tree representing an instance of
            *sym*1.
argument 4: first part of the difference list representing the
            string being parsed.
argument 5: second part of the difference list.

Operationally, if one is looking for a goal of *sym*1 and the grammatical category called is itself *sym*1, then the call terminates successfully using rule 1. In this case the second and third arguments, as well as the fourth and fifth are unified. If the goal in parsing is not *sym*1, rule 2 is used. A call to *link_rtc* (reflexive, transitive closure of the *link* relation; see below) determines whether this rule defining an *nt* can be used in parsing the goal *Goal*. If so, the predicate *goal* is called to successively recognize *sym*2, . . . , *symn* with derivation trees *Tree*2, . . . , *Treen* respectively. Finally, a call to *nt* tries to continue parsing toward the goal *Goal* with a portion of the derivation tree constructed from successful recognitions of *sym2*, *sym3*, etc., and the rest of the input string, to yield upon success *Treeout*, the portion of the derivation tree representing the *Goal*.

For example, the Prolog rules generated for the sample grammar's nonterminal *np* are:

```
np(Goal,TreeOut,Tree1,X0,X2) :-
link_rtc(s,Goal),
goal(vp,Tree2,X0,X1),
s(Goal,TreeOut,node(s,[Tree1,Tree2],'[]'),X1,X2).
```

np(np,Tree,Tree,X,X).

The process is begun by a call to *goal:*

?- goal(s,Tree,InputString,[]).

where *s* is the start symbol of the grammar, *Tree* is instantiated to the parse tree upon success, and *InputString* is instantiated to the string being parsed. For example:

?- goal(s,Tree,[pascal,stinks],[]),pretty(Tree).

produces the following reaction from Prolog:

```
0: s
  1: np
    2: [pascal]
  1: vp
    2: [stinks]
```

yes

The predicate *goal* is defined as follows:

```
goal(Goal,Tree,X,Z) :-
dict(Nonterminal,Tree1,X,Y),
link_rtc(Nonterminal,Goal),
P=..[Nonterminal,Goal,Tree,Tree1,Y,Z],
call(P).
```

In order to find an instance of *Goal* with derivation tree *Tree* from the string represented by the difference list of *X* and *Z*, we look up the first symbol of the input string, the difference between *X* and *Y*, in the dictionary *dict* and find it to be of category *Nonterminal,* and with derivation tree *Tree* 1; the relexive, transitive closure of the *link* relation verifies whether from recognizing a *Nonterminal* one may proceed to the *Goal*, and if so, constructs in *P* a term representing a call of *Nonterminal* with final goal *Goal* and final tree *Tree,* given that *Tree* 1 has been determined, and that the rest of the input string is represented by the difference of *Y* and *Z;* finally, a meta-call of *P* completes the definition of *goal.*

The *link* relation which is used to determine whether a grammar rule may be used in parsing toward a goal is determined as follows. Given a grammar rule of the form:

Nonterminal ::= Symbol,....

then

link(Symbol,Nonterminal)

may be asserted. The predicate *link_rtc* determines whether the two arguments are related by the reflexive, transitive closure of the *link* relation, i.e., whether the two arguments are identical, related directly by the *link* relation, or whether there is a grammatical symbol directly related to the first argument and which is in the *link_rtc* relation to the second.

Exercises

1. Define a Prolog predicate which computes for any binary relation its reflexive, transitive closure.
2. Use the techniques by which DCTGs are implemented to implement a translator from grammar rules to Prolog clauses as described above.
3. Modify your implementation so that grammar symbols may have extra arguments attached to them. See Matsumoto et al. (1983). □

Bibliographic Commentary for Part IV

Restriction Grammars implemented in Prolog are described in Hirschman and Puder (1982) and Hirschman (1987). Puzzle grammars are related in general philosophy to tree adjoining grammars developed by Joshi (1985). BUP, the bottom-up parser system, is described in Matsumoto et al. (1983). Earlier left-corner parsers and the use of what has come to be called the *link relation* are described in Ingerman (1966) and Abramson (1973). The use of the link relation, is however, usually ascribed to Pratt (1975).

PART V

LOGIC GRAMMARS AND CONCURRENCY

Chapter 13

Parsing with Commitment

1. Introduction

One of the most attractive features of logic programming is the possibility of executing logic programs in parallel. Recall that a logic program consists of a finite set of rules of the form:

$$A:-B_1, \ldots, B_n \qquad n \geq 0$$

where A and the B_i are atoms (see chapter 5). Given a query such as:

$$?- Q$$

the sequential method of solution of the query is to find some rule in the program whose head atom unifies with Q, say

$$Q':-R_1, \ldots, R_k$$

and replace the goal Q with the conjuction of goals $R'_1, \ldots, R'_k$ where the primes denote that the substitution derived from the unification of Q and Q' has been applied. If Q' had been a unit rule, then the query would have been solved. If not, the process continues by selecting some R'_j, finding a rule whose head unifies with that, replacing R'_j with the body of the selected rule under the derived substitution, and so on, until either all subgoals have been solved, or until no further progress can be made. The Prolog programming language (as opposed to pure logic programming) always selects the first goal, proceeds from left to right in seeking a solution, and backtracks when it runs into an impasse. Prolog considers the rules which unify with a goal atom in strictly textual order.

Pure logic programming, however, admits other possible strategies for solving a query. If R_i and R_j are independent in that they do not share any variables, then it is possible to try and solve both of these at once: this is called *Restricted AND parallelism* (see Gregory [1987]). If they share a variable, then it is possible to try to solve both at once provided that the incremental instantiation of the variable may be communicated between them: this is *Stream AND parallelism,* which allows one to express problems dealing with communicating processes executing concurrently. Other kinds of parallelism permitted by logic programming include: *OR parallelism* in which several rules whose heads unify with a goal atom are tried concurrently; and, *All-solutions AND parallelism* in which an attempt is made to find solutions to R_i and R_j which correspond to different solutions of the conjunction of goals. Research is being conducted in all of these areas, but it is the field of *Stream AND parallelism* which has resulted in a number of new logic programming languages, such as Concurrent Prolog, Parlog, and Guarded Horn Clauses, implementing a radically different goal solution strategy than Prolog's.

Given that there are possibilities of parallel solution of logic program goals, it is tempting to ask whether it is possible to extend notions of parallel execution to logic grammars. Analysis of independent clauses in a conjunction could be considered a form of *Restricted AND parallelism*. If we were looking up a word in a dictionary and it had several meanings and lexical functions, we could do the lookup on each of the meanings and functions concurrently, providing a form of *OR parallelism*. If we were trying to find all possible parses of a sentence we would be considering a form of *All-solutions AND parallelism*. If we considered nonindependent clauses of a sentence, we might think of the nonterminals which recognize these as processes which share some information, thus utilizing a kind of *Stream AND parallelism*. Research in these areas is just in its infancy: we shall describe below some work which has been done in trying to make use of the new languages (Concurrent Prolog, Parlog, Guarded Horn Clauses) for logic grammar applications. In particular we shall show that there is a kind of logic grammar whose rules can be easily compiled to clauses of these languages, but that this class of grammars, a generalization of LL(k) grammars, suggests that there are some limits of "expressiveness," though not of computing power, to programs written in theses languages. We shall also describe a more complicated method which permits general parsing of a restricted form of DCGs using a bottom-up stategy, but which has some limitations in the specification of "extra conditions," i.e., conditions enclosed between braces: $\{condition(X)\}$. This parsing method, however, could be used to find all possible parses of a sentence in the restricted formalism. Before we describe these two methods we must first consider in a bit more detail the notion of concurrency involved in *Stream AND parallelism*.

Sequential Prolog has been related to ordinary programming languages in the following way (Shapiro 1983): the list of rules with the same head atom corresponds to a procedure; solution of a goal corresponds to a procedure call; unification corresponds to the communication of parameters in a procedure call; and, sequential search with backtracking, simulating nondeterministic goal solution, corresponds to the ordinary execution mechanism of Von Neuman machines. In *Stream AND parallelism,* however, the comparision is made between any of the above mentioned languages and a system of concurrent, communicating processes: a conjunction of goals corresponds to a system of processes; a goal corresponds to a process; the state of the process is reflected in the values of goal arguments; communication between processes is by unification of common variables; synchronization of processes corresponds in Concurrent Prolog to suspending unification of variables which have been specified as "read-only" but not yet instantiated until some process which shares the variable has instantiated it (in Parlog, unification is suspended on variables which have been specified as being of "input mode" but not yet instantiated until some sharing process instantiates the variable); the computation of processes corresponds to indeterminate goal solution; and, failure of a process corresponds to the finite failure of an attempt to solve (reduce) a goal.

The form of a clause in these languages, ignoring differences in notation, is roughly :

$r(t_1, \ldots, t_k)$:- <guards> : <body>.

Both the guards and the body are a conjunction of goals. In attempting to evaluate a goal $r(p_1, \ldots, p_k)$, all clauses for the relation r are attempted in parallel. Here, attempted means matching the head of the clause and successful evaluation of the guards. From those clauses which are successfully attempted, one is selected, and the others are discarded. This is *don't care* or *committed* nondeterminism: the discarded calls are not cared about, or in other words, one is committed to a particular choice once made. In practice, the first attempt to succeed is chosen. In the languages mentioned there are also synchronization mechanisms for delaying calls to make sure that certain arguments are instantiated. In Concurrent Prolog this is done by annotating arguments as read-only; in Parlog, mode declarations specify which arguments are input or output arguments; in Guarded Horn Clauses, neither annotations nor mode declarations are used, but if during unification of the head or execution of the guards an attempt is made to bind a nonlocal variable, then execution suspends. During the attempt to evaluate a goal, any arguments, say in a guard, which are not instantiated when they should be, result in a suspension until the variable in question, shared by some other process, is more fully instantiated.

There is an obvious problem in parsing in a committed nondeterministic setting. From the productions that may be successfully attempted, the processor will select one, commit to it, and ignore all the others. This will obviously allow the derivation to continue one more step, but it may not allow the derivation to continue to a successful conclusion. For example, suppose we had the following productions for a nonterminal x and we were parsing in a setting of committed nondeterminism:

x ::= [].
x ::= a, b, c.

Suppose at some point in the parse, both productions were successfully attempted (assume empty guards) but that the processor had chosen the empty production to commit to. Even though the input may be parsed as an *a* followed by a *b* and a *c*, the wrong production (always applicable because of the empty right-hand side) will have been chosen and a parse will not be found.

Clearly, if there is any class of logic grammars for which there is a simple direct translation of grammar rules to concurrent logic program clauses in a setting of committed nondeterminism, then that production (clause) must always be selected which will allow a derivation to continue to a successful conclusion if one exists This will happen if at any time at most one production can be used to continue a derivation. Fortunately, there is a subclass of context-free grammars, the LL(k) grammars, which provides a model for such a class of logic grammars. The class of LL(k) grammars consists of those unambigous context-free grammars in which input is parsed top-down from left to right with k-symbol lookahead. The lookahead enables one to uniquely determine which production is to be used in continuing a parse. If no production is applicable, then the input string is not in the language generated by the grammar. This class is deterministic in the sense that it can be accepted by a deterministic pushdown automaton.

We shall use the following LL(1) grammar, first showing how deterministic grammars may be compiled as sequential logic programs and then, generalizing, compiled to concurrent logic programs with ''don't care'' nondeterminism (note the paradox: deterministic grammars compile easily into don't care nondeterministic logic programs!).

1.1. Sample Grammar
e ::= t,e_prime.

e_prime ::= "+",t,e_prime.
e_prime ::= [].

t ::= f,t_prime.

t_prime ::= "*",f,t_prime.
t_prime ::= [].

f ::= "a".
f ::= "(",e,")".

See the end of this section for comments on nondeterministic parsing.

1.2. The One Character Lookahead Relation
The following unit clauses define the one character lookahead relation for the sample grammar. The first argument to the predicate *lookahead* is a production, and the second is a list of characters that permit use of that production in a derivation. For example, the production *e::=t,e_prime* may be used if searching for an *e* and if the first unused character in the input string, the lookahead character, is either an *a* or a (.

lookahead((e::=t,e_prime),"a(").
lookahead((e_prime::="+",t,e_prime),"+").
lookahead((e_prime::=[]),")?").
lookahead((t::=f,t_prime),"a(").
lookahead((t_prime::="*",f,t_prime),"*").
lookahead((t_prime::=[]),"+)?").
lookahead((f::="a"),"a").
lookahead((f::="(",e,")"),"(").

The *?* is used to mark the end of the input string. The *lookahead* predicate may be calculated following an algorithm given in Aho and Ullman (1972, 1977) and is easily specified in Prolog (although some version of *setof* must be used).

2. Compilation to Sequential Logic Programs

Although our principal motivation is to find a class of concurrent logic grammars, we begin with the compilation of deterministic grammars to sequential logic program clauses: the sequential case is itself interesting and gives the foundation of the method to be developed for the concurrent case.

In compiling LL(k) grammars to sequential logic programs we would like to take advantage of the determinacy of production use in a derivation. We shall do so by treating the lookahead examination as a guard on the use of a clause compiled from a production. The logic program clauses generated from the above grammar will each have as their first goal a call to a predicate called *ll_guard*. This is a predicate of arity 3: the first argument is the original production used to index into the lookahead predicate; the second argument is treated as a node in the tree representation of the derivation and is a function symbol of the form *guard(X)*, recording the lookahead string *X;* the last two arguments represent the input string as a difference list. For the LL(1) case, the *ll_guard* predicate is specified as:

```
ll_guard(LookAhead,guard(X),[X|Xs]) :-
  lookahead(LookAhead,List),
  member(X,List),!.
```

This does not use up any characters in the input string but merely examines them. The cut, once the lookahead *List* has been accessed, and the first character of the input string has been shown to be a member of that list, is used to ensure that there will be no backup in trying to use any other productions. (A clever compiler might avoid generating choice points which would not be used.)

Here follows the set of clauses generated for the above sample grammar. The call to *ll_guard* is automatically inserted by the grammar compiler. The third argument to the function symbol *node* represents an empty set of semantic rules.

```
e(node(e,[Guard,T,E_prime],[]),S1,S3) :-
  ll_guard((e::=t,e_prime),Guard,S1),
  t(T,S1,S2),
  e_prime(E_prime,S2,S3).

e_prime(node(e_prime,[Guard,[+],T,E_prime],[]),S1,S4) :-
  ll_guard((e_prime::=[+],t,e_prime),Guard,S1),
  c(S1,+,S2),
  t(T,S2,S3),
  e_prime(E_prime,S3,S4).

e_prime(node(e_prime,[Guard,[]],[]),S1,S2) :-
  ll_guard((e_prime::=[]),Guard,S1),
```

```
     S1=S2.

t(node(t,[Guard,F,T_prime],[]),S1,S3) :-
   ll_guard((t::=f,t_prime),Guard,S1),
   f(F,S1,S2),
   t_prime(T_prime,S2,S3).

t_prime(node(t_prime,[Guard,[*],F,T_prime],[]),S1,S4) :-
   ll_guard((t_prime::=[*],f,t_prime),Guard,S1),
   c(S1,*,S2),
   f(F,S2,S3),
   t_prime(T_prime,S3,S4).

t_prime(node(t_prime,[Guard,[]],[]),S1,S2) :-
   ll_guard((t_prime::=[]),Guard,S1),
   S1=S2.

f(node(f,[Guard,[a]],[]),S1,S2) :-
   ll_guard((f::=[a]),Guard,S1),
   c(S1,a,S2).

f(node(f,[Guard,['('],E,[')']],[]),S1,S4) :-
   ll_guard((f::=['('],e,[')']),Guard,S1),
   c(S1,'(',S2),
   e(E,S2,S3),
   c(S3,')',S4).
```

The predicate *c* is used to absorb a single terminal symbol:

```
c([X|Y],X,Y).
```

The controlling predicate *e* appends the endmarker, in this case, *?*, calls the starting nonterminal of the grammar, and pretty prints the result:

```
e(Source) :-
   append(Source,[?],EndMarked),
   e(Guard,EndMarked,[?]),
   pretty(Guard).
```

For example, a call of *e("a*a")* yields:

```
e
  guard(a)
  t
    guard(a)
    f
```

```
    guard(a)
    [a]
  t_prime
    guard(*)
    [*]
    f
      guard(a)
      [a]
    t_prime
      guard(?)
      []
  e_prime
    guard(?)
    []
```

3. Compilation to Concurrent Logic Program Clauses

We shall illustrate the compilation of a deterministic grammar to a (don't care) concurrent logic program using the language Concurrent Prolog as a target; compilation to Parlog and GHC is similar, and we shall comment on this below. The basic idea is to turn the predicate *ll_guard* into a true guard on the generated clause and each nonterminal into a concurrent process. An attempt is made to reduce the generated clause only if the guard succeeds. The processes corresponding to nonterminals must be synchronized so that there is, in the LL(1) case, a character in the input string against which a guard may succeed or fail. The synchronization is accomplished by annotating the first of the two hidden arguments with a *?*, the read-only annotation. If the input string is not yet sufficiently instantiated, the process delays until an input character has appeared. Here are the generated Concurrent Prolog clauses for our sample grammar. The commit operator is indicated by a ;.

```
e(node(e,[Guard,T,E_prime],[]),S1,S3) :-
  ll_Guard((e::=t,e_prime),Guard,S1);
  t(T,S1?,S2),
  e_prime(E_prime,S2?,S3).

e_prime(node(e_prime,[Guard,[]],[]),S1,S2) :-
  ll_Guard((e_prime::=[]),Guard,S1);
  S1? = S2.

e_prime(node(e_prime,[Guard,[+],T,E_prime],[]),S1,S4) :-
  ll_Guard((e_prime::=[+],t,e_prime),Guard,S1);
  c(S1?,+,S2),
  t(T,S2?,S3),
  e_prime(E_prime,S3?,S4).
```

```
t(node(t,[Guard,F,T_prime],[]),S1,S3) :-
  ll_Guard((t::=f,t_prime),Guard,S1);
  f(F,S1?,S2),
  t_prime(T_prime,S2?,S3).

t_prime(node(t_prime,[Guard,[]],[]),S1,S2) :-
  ll_Guard((t_prime::=[]),Guard,S1);
  S1? = S2.

t_prime(node(t_prime,[Guard,[*],F,T_prime],[]),S1,S4) :-
  ll_Guard((t_prime::=[*],f,t_prime),Guard,S1);
  c(S1?,*,S2),
  f(F,S2?,S3),
  t_prime(T_prime,S3?,S4).

f(node(f,[Guard,['('],E,[')']]],[]),S1,S4) :-
  ll_Guard((f::=['('],e,[')']),Guard,S1);
  c(S1?,'(',S2),
  e(E,S2?,S3),
  c(S3?,')',S4).

f(node(f,[Guard,[a]],[]),S1,S2) :-
  ll_Guard((f::=[a]),Guard,S1);
  c(S1?,a,S2).
```

We use the following definition of *member* in the Concurrent Prolog setting:

```
member(X,[Y|_]) :-
  X = Y ; true.

member(X,[Y|Z]) :-
  dif(X , Y);
  member(X,Z).
```

The definition of the *ll_guard* predicate must be changed slightly since it examines the input string: it delays until there is a character in the input stream and the lookahead *List* has been supplied:

```
ll_guard(LookAhead,guard(X),[X|Xs]) :-
  lookahead(LookAhead,List),
  member(X?,List?).
```

The controlling predicate now calls on the Concurrent Prolog interpreter to solve the goal *e,* with the endmarked input string, yielding if successful, the derivation tree *T:*

```
e(Source) :-
 append(Source,"?",EndMarked),
 solve(e(T,EndMarked,"?")),pretty(T).
```

In the case of Parlog, the compiler from grammars to Parlog clauses would have to annotate the processes corresponding to nonterminals with mode declarations which would insure that the next to the last argument is an input variable. The predicate *ll_guard* would act as a guard to the generated Parlog clauses. This example has in fact been converted to Parlog by S. Gregory (personal communication). Conversion of this technique to Guarded Horn Clauses should not be difficult.

4. Generalized Deterministic Grammars

We have so far shown how LL(k) grammars could be compiled directly into either sequential or don't care nondeterministic logic programming languages. The class of LL(k) languages is in some respects a restrictive one: it does not include all context-free grammars, for example. Thus, one could not take an arbitrary context-free grammar and transform it into an LL(k) grammar and then generate an efficient logic program (efficient in the sense of not requiring backtracking). In practice, however, many languages (probably most programming languages) can be formulated using LL(k) grammars. It is fairly likely that any language (presumably, for convenience to the user, a fairly restricted subset of natural lanaguage) which might be used as a command language to a logic operating system could be specified by an LL(k) grammar for some small value of k. One would then be able to use the hardware of a concurrent logic machine to handle the necessary grammatical processing directly rather than relying on an attached sequential grammar processor for this task.

There are, however, some obvious generalizations of the techniques displayed above which get out of the restrictive LL(k) class.

Firstly, the guards may be generalized to do more than look at a certain number of characters of the input stream. Grammar productions could be written in the form:

nonterminal ::= <guards>: <right-hand side>

where the nonterminal expands to the right-hand side only if the guards are successfully evaluated. It would be up to the grammar writer to provide specifications of the guards so that the wrong production is not selected in the committed non-deterministic setting.

Secondly, the nonterminal symbols may be allowed, as in the case of Definite Clause Grammars or Definite Clause Translation Grammars, to have more arguments than just those automatically added by the compiler from grammar rules to logic programming clauses. This certainly takes the grammar rules out of the very restrictive LL(k) class, and even, as is well known from the DCG and DCTG experience, out of the class of context-free grammars.

Thirdly, the right-hand side of extended grammar rules may also include communication with concurrent processes other than ones corresponding to nonterminal
symbols. As in the case of DCGs and DCTGs, one might use the notation in the
right-hand part of a grammar rule:

{ concurrent_process(A, . . . , Z) }

to specify that some concurrent process with shared variables A to Z must be successfully reduced for parsing to succeed.

One should also note that in the concurrent setting derivations may be of infinite
length. The guards above are used to determine whether a derivation may be con-
tinued or not: they do not enforce any restrictions on the length of input. Thus, one
may think of the generalized grammar rules as allowing one to do grammatical
processing on streams rather than on finite strings. In this view, the grammar rules
provide a notation for operating on what might be termed a "hidden stream": it is a
mechanical task to generate the concurrent logic program clauses which make that
stream explicit as above in the simple LL(1) case.

5. Related Work

An analogy can be made between SLD-resolution over Horn clause programs and
context-free derivations over context-free grammars. In place of grammar rules
one has a program of Horn clauses; instead of replacing a nonterminal by the
right-hand side of a grammar rule whose left-hand side is that nonterminal, one
seeks to unify a goal with the head of a clause, and if successful replace that goal
either by the empty clause or by the body of the clause using the substitution
derived from unification to instantiate variables. In the context-free grammar
situation one tries to derive a sentential form without nonterminals; in SLD-
resolution one tries to remove all subgoals, deriving the empty clause. Thus,
SLD-resolution over Horn clause programs generalizes context-free derivations.

In this section we have drawn an analogy between LL(k) grammars, a subclass of
context-free grammars and commited choice non-deterministic concurrent logic
programming. LL(k) grammars constitute a proper subclass of context-free grammars that can be parsed efficiently. The drawback to this class is that the grammars
of the class are viewed as being less "natural" and less "expressive" than full
context-free grammars. Given the analogy drawn between LL(k) grammars and
committed choice concurrent programming languages, one hopes that the lack of
"naturalness" and "expressiveness" characteristic of the grammars does not
carry over to the programming languages. If it does, one might wish to investigate
specifying problems in full and/or parallel logic and use some heuristic program
transformation techniques to derive efficient, but possibly less "natural," committed choice concurrent programs. Note that committed choice concurrent programming languages, as well as generalized LL(k) grammars for such languages, have
the full computing power of a Turing machine. The concepts *natural* and *expres-*
sive hence are intuitive and must be placed within quotation marks.

Other approaches may be taken to parsing in languages such as Parlog, Concurrent Prolog, and Guarded Horn Clauses. One could simply avoid the problem and drop into Prolog, making use of known classes of logic grammars for parsing; if all possible parses of a sentence were required, one could make use of various "all solutions" predicates for gathering the parses into a list. This method, although effective, is not very interesting as far as exploitation of concurrent logic programming languages is concerned.

The approach taken by Matsumoto (described in the next section) with respect to parsing in a concurrent setting is an alternative to the generalized LL(k) grammars and is more general, but posssibly less efficient. Matsumoto's approach is to allow nondeterministic grammars and use a parsing method related to Chart Parsing and Earley Parsing. Potentially, all possible parses may be gathered and merged into a list of parses. This seems quite suited to nondeterministic natural language parsing but may be unnecessarily powerful when used with deterministic formal languages. Also, there are problems in that $\{condition(X)\}$ calls can be evaluated only once, and hence must be restricted to deterministic conditions.

Having said that nondeterministic parsing may be more suitable for natural language parsing in general, we still think that deterministic concurrent parsing may be applied to natural language parsing in some cases. Marcus has reported considerable success with a deterministic bottom up parser which is essentially an LR(3) parser. It is tempting to speculate that a top-down analogue of his parser can be as successful.

On a less speculative level, one would like to have the grammatical processes in the concurrent setting as efficient and as inexpensive as possible. For much of the time the process corresponding to a nonterminal may be inactive, coming alive only when some input had arrived on its input stream. These processes could presumably be efficiently implemented by having them do a busy wait or be blocked until activated.

6. Parallel Parsing for Natural Language Analysis

The method we are going to describe in this section was developed by Matsumoto and is based on the left corner parsing method described in the chapter on BUP – the bottom-up parser embedded in Prolog. The examples will be written in Parlog, but can easily be modified for Flat Concurrent Prolog and Guarded Horn Clauses. We follow Matsumoto's presentation closely in this section.

6.1. Sample Grammar

```
s   ::= np, id1 vp.
np ::= det, id2 noun.
np ::= det, id3 noun, id4 relc.
relc ::= [that], id5 s.
vp ::= verb.
vp ::= verb, id6 np.
det ::= [the].
```

```
noun ::= [man].
noun ::= [woman].
verb ::= [loves].
verb ::= [walks].
```

The positions marked by *idn* are introduced to mark positions between grammatical symbols as in Chart Parsing. Corresponding to the above grammar rules, the following Prolog clauses can be generated to implement a left-corner parser for the language generated by the grammar.

```
np(X,id1(X)).
vp(id1(X),Y) :- s(X,Y).
det(X,id2(X)).
noun(id2(X),Y) :- np(X,Y).
det(X,id3(X)).
noun(id3(X),id4(X)).
relc(id4(X),Y) :- np(X,Y).
that(X,id5(X)).
s(id5(X),Y) :- relc(X,Y).
verb(X,Y) :- vp(X,Y).
verb(X,id6(X)).
np(id6(X),Y) :- vp(X,Y).
the(X,Y) :- det(X,Y).
man(X,Y) :- noun(X,Y).
woman(X,Y) :- noun(X,Y).
loves(X,Y) :- verb(X,Y).
walks(X,Y) :- verb(X,Y).
```

Exercises

1. Write Prolog predicates to generate the *idn*-marked rules from context-free grammar rules.

2. Write Prolog predicates to generate the Prolog clauses from the result of the last exercise. Hint: see the chapter on BUP.

3. Verify that the Prolog clauses generated constitute a left-corner parser for the grammar. For the sentence *The woman loves* the following goal should be solved:

   the(begin,X),woman(X,Y),loves(Y,end).

   where *begin* and *end* are symbols that mark the beginning and end of a sentence. One more rule will be necessary for parsing:

   s(X,Y) :- X==begin,Y==end.

Explain. □

6.2. Parallel Parsing Method

The clauses in the above Prolog program are of two kinds: one kind has a variable for its first argument; the other has a structured item (function symbol) for its first argument. The clauses of the first type, called variable type clauses, start to construct a new piece of the derivation tree in which the clause is the left-corner constituent of the new tree structure being constructed. The clauses of the second type are called structure type clauses. A structure type clause tries to unify the first element of the received sequence against its first argument, and if it can do so, either makes another Prolog call, or builds more data structure.

Many computations are repeated during the course of parsing as described. This is because variable type clauses try to put an identifier on the received data of the history of the parse without considering the content of the history. To avoid repetitions, histories of the parse so far are combined into a set, and variable type clauses are applied to the set of possible parses so far. The structured type clauses can then receive a set of histories and try to see which of them (if any) may be extended by use of that structured clause.

Given a nonterminal, its variable type clauses and structured type clauses can be run in parallel, each receiving the same data. If the set of histories is treated as a stream of data, the variable type clauses generate the stream, and the structured type clauses modifies a stream coming to it.

Matsumoto gives as an example the following Parlog clauses which implement these considerations. The Parlog clauses generated from variable type clauses of the Prolog program will end in *v,* those generated from structured type clauses will end in *s.* Corresponding to the clauses for *det:*

```
det(X,id2(X)).
det(X,id3(X)).
```

are the Parlog clauses:

```
mode det_v(?,^).
det_v(X,[id2(X),id3(X)]).
```

The ? indicates an input argument, the ^ indicates an output argument. Corresponding to the structured type clauses for noun:

```
noun(id2(X),Y) :- np(X,Y).
noun(id3(X),id4(X)).
```

are the Parlog clauses:

```
mode noun_s(?,^).
noun_s([],[]).
```

```
noun_s([id2(X)IT], Y) :- I
 np(X,Y2),
 noun_s(T,Y1),
 merge(Y1,Y2,Y).
noun_s([id3(X)IT],[id4(X)IY] ) :- I
 noun_s(T,Y);
noun_s([_IX],Y) :- I
 noun_s(X,Y).
```

In the Parlog variable clauses, the first argument, using the same variable name as in the Prolog clauses, assumes that the input is a stream of histories, which is represented as a list of streams headed by some identifier. These variable type clauses collect all left-corner branches starting at their own position. In the case of variable type clauses which have a variable also in the second argument (derived from grammar rules in which there is only one symbol on the right-hand side), a special treatment is possible. For example:

```
verb(X,Y) :- vp(X,Y).
verb(X,id6(X)).
```

can be merged into the following single Parlog clause:

```
mode verb_v(?,^).
verb_v(X,[id6(X)IY]) :- I
 vp(X,Y).
```

Outputs of more than one such clause get merged together.

The structured type clauses now constitute a set of OR-processes each of which tries to extend a set of histories by a particular identifier. In the definitions of *noun_s*, the first clause takes care of the case of an empty stream. The second clause deals with the situation when a history is headed by *id2:* a noun phrase has been completely recognized and is reduced to *np*, and another call to *noun_s* to deal with remaining data in the stream (the next *id* 2). The third Parlog clause deals with the situation when a noun modifies a partly completed tree, but the result is not yet complete. The last Parlog clause (preceded by a ; indicating that it is not to be used except when preceding clauses fail to apply) is used when the other clauses of *noun_s* have not been able to deal with the first item of the stream. It simply discards the first element of the stream and calls itself to operate on any remaining data.

For every nonterminal symbol in the grammar, a variable type process and a structured type process is generated:

```
mode noun(?,^).
noun(X,Y) :- I
 noun_v(X,Y_v),
 noun_s(X,Y_s),
```

merge(Y_v,Y_s,Y).

For the nonterminal symbol *s*, the following Parlog clause is introduced for a structured clause:

s_s(X,[end]) :- X==[begin]|true.

and the goal of parsing a sentence is phrased as:

the([begin],X),woman(X,Y),loves(Y,Z),finish(Z).

The process *s_s* inserts the *end* which indicates when a sentence is complete. The user defines a process called *finish* which accepts the stream produced by the last word of the sentence, makes sure it contains the word *end*, and then makes whatever use of the results of the parse that is desired.

The correspondence between Chart Parsing and this kind of parallel parsing is indicated by Matsumoto as being rather simple: inactive edges of chart parsing correspond to calls of nonterminal symbols, and the active edges of chart parsing correspond to the data structures passed to the nonterminal symbols as arguments. The first and second rules of chart parsing correspond to the variable type and structured type clauses respectively. Since Chart Parsing generalizes Earley's parsing algorithm, this method of Matsumoto's also includes Earley's algorithm as a special case.

One difficulty with this parsing method is that the so-called *extra conditions*, constraints of the form {*condition* (*X*)}, must be restricted to be deterministic and must not instantiate variables in the bodies of the grammar rules. The constraints must be deterministic because they are evaluated only once in the system. Variables in grammar rules must not be instantiated because, in case of ambiguities which spawn many processes, the same variables which should be in different environments are treated by the system as being in a single environment. A proper treatment of such variables would require copying or renaming of the variables, costly in both time and space.

Exercises

1. Write Prolog predicates to generate the variable type and structured type clauses from a grammar.

2. The explicit merge processes above may be removed by using difference lists. How is this done?

3. Given the results of the previous exercise, the grammar rules can be made to look like DCG rules. How is this done?

4. Make explicit the connection between Chart Parsing of a grammar and the Parlog clauses generated for a grammar by this method.□

Bibliographic Commentary for Part V

The three principal committed choice languages, Concurrent Prolog, Parlog, and Guarded Horn Clauses were introduced in Shapiro (1983), Clark and Gregory (1983) and Ueda (1985), respectively. Parlog and its implementation are thoroughly discussed in Gregory (1987), while the two volumes of Shapiro (1988) contain many important papers on Concurrent Prolog and it's applications. The series of technical reports from ICOT, Tokyo, contains much information related to Guarded Horn Clauses and their applications. There is a great deal in common among these three languages, but also subtle differences. See Crammond (1986) for discussion of a common execution model for these languages.

The best sources for further information on the other kinds of parallelism are the proceedings of the international logic programming conferences and symposia.

LL(k) grammars were introduced by Knuth (1971). The hand-coding of LL(k) grammars into Prolog clauses was very briefly mentioned in Stabler (1983). Matsumoto's parallel parsing method appears in Matsumoto (1986). Earley's general context-free parsing method is described in Earley (1970). Chart parsing can be traced back to Kaplan (1971) and Kay (1976).

APPENDIX I[1]

INPUT/OUTPUT IN LOGIC GRAMMARS

1. Input of a Sentence

The following section of code allows for the input of a whole sentence to the parser rather than the usual list of atoms, i.e. a list of words separated by commas.

```
phh([]) :- nl.
phh([H|T]) :- write(H),tab(1),phh(T).

hello :- phh(['listening']),
         read_in(S),write(S),nl,
         sent(X,S,[]),
         nl, phh(['meaning']),
         write(X),nl,nl.

read_in([W|W4]) :- get0(C),readword(C,W,C1),restsent(W,C1,W4).

restsent(W,_,[]) :- lastword(W),!.
restsent(W,C,[W1|W4]) :- readword(C,W1,C1), restsent(W1,C1,W4).

readword(C,W,C1) :- single_character(C), ! , name(W,[C]),get0(C1).
readword(C,W,C2) :-
    in_word(C,N),
    ! , get0(C1),
    restword(C1,C4,C2),
    name(W,[N|C4]).
readword(C,W,C2) :- get0(C1),readword(C1,W,C2).

restword(C,[N|C4],C2) :- in_word(C,N), ! , get0(C1),
 restword(C1,C4,C2).
 restword(C,[],C).

single_character(63).
single_character(39).
single_character(46).

in_word(C,C) :- C>96,123>C.
```

---

[1] N.B. We believe a student collaborated in this section, however, we unfortunately cannot recall his or her name. If you happen to be that student, please identify yourself for appropriate credit in future editions.

```
in_word(C,L) :- C>64,C<91, L is 32 + C.
in_word(C,C) :- C > 47, 58 > C.
in_word(45,45).
lastword('.').
lastword('?').
```

The sentence has to be terminated by a . or a ?.

This program presumes that a compiled grammar is already present. Thus the program knows the meaning of sent(X,S,[]).

The clause *read_in* translates a sentence like *I am happy* to the form [i, am, happy, .] which is required by the grammar.

The usage is ?- *hello*.

The system answers */listening*.

At this point the desired sentence is typed in with a return at the end and the system responds with an echo of the sentence in list form, the word *meaning* on a new line, and then gives the representation on the final line.

Automating Calls for Parsing and Generating

Rather than carry the output empty string explicitly as in the above call to *sent*, we next define the predicates *parse* and *generate*, which were used in section 6 of chapter 1.

```
parse(F,S):- tr(process(F,S),P), call(P).
generate(F,S):- tr(process(F,S),P),call(P).

tr(process(F,S),P):- F=..[N,A], concat(A,[S,[]],A1),
          P=..[N|A1].

concat(X,Y,[X|Y]):- var(X),!.
concat([],Y,Y).
concat([X|Y],Z,[X|W]):-concat(Y,Z,W).
concat(X,Y,[X|Y]).
```


Exercises

1. Modify the procedure *hello* above so that it uses *parse* or *generate* according to the user's choice. □

APPENDIX II

IMPLEMENTATION OF LOGIC GRAMMAR FORMALISMS

1. SYNAL: Compiling Discontinuous Grammars into Prolog

The following implementation of discontinuous grammars was designed by V. Dahl and coded by M. McCord during joint work on coordination, which was later superceded by their work on Modifier Structure Grammars, described in chapter 8.

This program compiles grammar rules in the --> notation into a Prolog program. It operates in the environment of C-Prolog and in fact uses its --> operator and its primitive *expand_term* for compiling discontinuous grammar rules into Prolog. This operator and the corresponding clauses for *expand_term* are discussed, for instance, in appendix A of Clocksin and Mellish (1981); *expand_term* is named "translate" in that reference.

```
synal((A,B --> C),Clause) :- !,
 expand_term((c_nonterm-->C),CClause),
 expand_term((b_nonterm-->B),BClause),
 clauseparts(CClause,CHead,CBody),
 clauseparts(BClause,BHead,BBody),
 CHead=..[c_nonterm,X,Z],
 BHead=..[b_nonterm,Y,Z],
 A=..[Pred|Args],
 concaten(Args,[X,Y],NewArgs),
 NewA=..[Pred|NewArgs],
 combine(CBody,BBody,Body),
 formclause(NewA,Body,Clause).

synal(C1,C2) :- expand_term(C1,C2).

clauseparts((Head:-Body),Head,Body) :- !.
clauseparts(Head,Head,true).

formclause(Head,true,Head) :- !.
formclause(Head,Body,(Head :- Body)).

combine(true,B,B) :- !.
combine(A,true,A) :- !.
combine(A,B,(A,B)).

concaten([X|L],M,[X|N]) :- concaten(L,M,N).
concaten([],L,L).
```

```
skip([])-->[].
skip([Word|Rest])-->[Word],skip(Rest).

sa:-read(Rule),synal(Rule,Rule1),assertz(Rule1).
sa(File):-see(File),start,seen.
sa(File,Filout):-see(File),tell(Filout),start,seen,told.

start:-read(Rule),process(Rule).

process(Rule):-Rule=done,!.
process(Rule):-synal(Rule,Rule1),assertz(Rule1),pp(Rule1),nl,start.

pp((A :- B)) :- !,write(A), write(' :'), write('-'), nl, ppr(B), write('.'), nl.
pp(A) :- write(A), write('.'), nl.

ppr((B,C)) :- !, write('   '), write(B),write(',') ,nl, ppr(C).
ppr(B) :- write('   '), write(B).
```

The program operates on rules contained in a file X. It is invoked by the query:

> :- sa(X).

The resulting Prolog rules are asserted. Alternately, rules in X can be read and their Prolog counterparts be stored in a file Y by using the query:

> :- sa(X,Y).

Rules are assumed normalized – i.e., rules with more than one left-hand side symbol must have only one nonterminal (the leading one) on the left-hand side. To ensure this, any nonterminal nt in a left-hand side, other than the leading symbol, is replaced by the pseudo-terminal $[nt]$, and a rule $nt \longrightarrow [nt]$ is added to the set of rules.

2. A Short Interpreter for Static Discontinuity Grammars

Interpreters are not the most efficient implementations for grammar formalisms, but they are very useful for testing various parsing strategies. We present one of the versions we have been experimenting with, consisting of eleven Prolog rules which interpret discontinuous productions in SDGs by applying each applicable subrule as soon as possible. It can be viewed as a pseudo-parallel interpreter, in that if all subrules are immediately applicable, they will be applied at once, achieving a parallel effect.

The main procedure is **parse(String,State,Pending_subrules)**, where **String** is the list of words of in a presumed sentence in the language defined by the input grammar, **State** is the current parsing state, and **Pending_subrules** is a list of pending subrules from a type-2 production. The first call must initialize **State** to a list containing the initial suymbol, and **Pending_subrules** to *[]*.

Type-1 productions are stored in the form:

    rule(Head,Bodylist).

and type-2 productions, in the form:

    rule([[Head1,Bodylist1],...,[Headn,Bodylistn]]).

The grammar must have stopping clauses before recursive ones, to avoid looping.

The first rule for *parse* is the terminating one; the second recognizes a terminal symbol; the third applies an applicable pending subrule left from a previous partial application of a type-2 production. The fourth rule applies a matching type-2 production by applying its first subrule and concatenating all others into the Pending_subrules list.

Notice that different strategies can be easily adopted through small changes to this interpreter. For instance, the *getsubrule* predicate, which presently chooses subrules in order unless otherwise provoked through backtracking, can be changed to randomly choose any subrule, or to choose them in a predefined order.

```
parse([],[],[]).

parse([T1|T],[T1|S],P):- parse(T,S,P).

parse(S,G,P):- getsubrule(P,[H|B],P1),   /* if any subrule can be */
          break(G,F,H,L),                /* applied, apply it */
          concat(F,B,F1),
          concat(F1,L,G1),
          parse(S,G1,P1).
```

```
parse(S,[H|G],P):- rule([H,B],Sr),   /* apply a type-2 rule, */
                concat(B,G,G1),   /* append waiting */
                concat(Sr,P,P1),  /* subrules on the pending list */
                parse(S,G1,P1).

parse(S,[H|G],P):- rule(H,B),              /* apply a type-1 rule  */
       concat(B,G,G1),
       parse(S,G1,P).

/* getsubrule(List of pending rules, Subrule chosen, */
/*                    Newlist of pending rules)*/

getsubrule([R|P],R,P).       /* gets a subrule out of the pending list*/

getsubrule([R1|P],R,[R1|P1]):- getsubrule(P,R,P1).

break([H|L],[],H,L).
break([H1|L],[H1|F],H,L1):- break(L,F,H,L1).

concat([],Y,Y).
concat([X|R],Y,[X|Z]):- concat(R,Y,Z).
```

3. Implementations of DCTGs

3.1. An Interpreter for Definite Clause Translation Grammars

For the Definite Clause Grammar Interpreter we require the following operator definitions. These define the various operators used in the DCTG notation.

    :- op(650,yfx,^^).
    :- op(1150,xfx,::=).
    :- op(1175,xfx,<:>).
    :- op(1150,xfx,::-).

The numbers specifying binding levels are of course implementation dependent.

We also assume in the interpreter below, that the right hand sides of DCTG rules are preprocessed to a list, i.e., a DCTG rule such as:

    a ::= b,c,{d}<:>Semantics.

is represented internally as:

    a ::= [b,c,{d}]<:>Semantics.

In the following clauses, the first argument represents the current grammatical symbol in a DCTG rule, the second argument represents the subtree being constructed for that symbol, the third and fourth arguments represent the difference list containing the input (or output) string of symbols.

1. parse([First|Rest],[FirstTree|RestTree],S0,SX) :-
        parse1(First,FirstTree,S0,SY),
        parse(Rest,RestTree,SY,SX).

2. parse([],[],SX,SX).

3. parse1([Terminal],[Terminal],SX,SY) :-
        connect(SX,Terminal,SY).

4. parse1({Prolog}, call(Prolog), SX, SX) :-
        call(Prolog).

5. parse1((First^^FirstTree),FirstTree,S0,SX) :-
        (First ::= BodyList <:> Semantics),
        parse(BodyList,Subtrees,S0,SX),
        FirstTree = node(First,Subtrees,Semantics).

6. parse1(First,node(First,Subtrees,Semantics),S0,SX) :-
        (First ::= BodyList <:> Semantics),

parse(BodyList,Subtrees,S0,SX).

7.   connect([A|B],A,B).

Clause 1 interprets a list of grammatical symbols [*First* |*Rest*] which use up the input string between points *S*0 and *SX*, producing the list of trees [*FirstTree* |*RestTree*]. *parse* 1 is called to interpret the *First* symbol between points *S*0 and *SY* and the *FirstTree* is produced. Then *parse* is called recursively on *Rest* to produce *RestTree* and to use up symbols between *SX* and *SY*.

Clause 2 interprets the empty list of grammatical symbols, producing an empty tree, and the input and output lists are unified.

Clause 3 interprets a single terminal symbol between *SX* and *SY*. The subtree produced is of the form [*Terminal*].

Clause 4 interprets a call to a Prolog query: no symbols of the input string are used up, but a node is included in the derivation tree to indicate the call of a Prolog predicate.

Clause 5 interprets a single nonterminal symbol which is decorated with the name of a subtree *First*^^*FirstTree* producing *FirstTree* and using symbols between *S*0 and *SX*. A grammar rule must be found with *First* on the left-hand side and with the right-hand side *BodyList* <:>*Semantics*. A recursive call to parse interprets the symbols in BodyList, using symbols between *S*0 and *SX*, and producing the list of *Subtrees*. *FirstTree* is instantiated to a node of the form *node(First,Subtrees,Semantics)*.

Clause 6 is similar to clause 5, except that the subtree corresponding to First is not named, implying that this subtree does not take part in any semantic action.

Clause 7 identifies a terminal symbol *A* as the first symbol in the input list represented as a difference list between [*A* |*B*] and *B*.

8.   [(Head::-Body) |_]^^ Head :-
         Body.

9.   [Head |_]^^ Head.

10.   [_|Rules]^^ Head :-
         Rules^^ Head.

Clauses 8, 9, and 10 specify how semantic rules are evaluated.

Clause 8 is used when a semantic goal instantiating *Head* matches the head of a semantic rule *Head* ::-*Body* : the *Body* is simply evaluated.

Clause 9 is used for a unit semantic rule.

Clause 10 searches the remaining semantic *Rules* for one whose head matches *Head*.

3.2. Compiling Definite Clause Translation Grammars to Prolog

In this section we shall describe how one class of logic grammars (DCTGs) is compiled into a logic program. This will serve as a guide to the specifications of how other logic grammar formalisms may be compiled into logic programs, and also as a pattern for the reader to follow in devising compilers for any new formalisms. The ambitious reader may wish to treat the compilation of a logic grammar to a logic program as an exercise, and not read the following section until the exercise has been completed.

The reader is referred to the section on Definite Clause Translation Grammars for a specification of what may enter into a DCTG rule. We presume familiarity with that description in what follows.

Operator Specifications

We require the following operators in specifying DCTG rules:

```
<:>    ::=    ^^    ::-
```

These may be specified in Prolog (Edinburgh-C) as follows:

```
:- op(650,yfx,^^).
:- op(1150,xfx,::=).
:- op(1175,xfx,<:>).
:- op(1150,xfx,::-).
```

The reader is referred to a suitable Prolog manual or textbook for the meaning of the magic numbers and symbols. Briefly, they establish the priority and binding properties of these new infix operators for the Prolog syntax analyzer, permitting them to be used in conjunction with existing ones.

The Predicate Translate_Rule

We remind the reader that the general form of a DCTG rule is:

Syntax <:> Semantics

with the syntactic portion of the rule partitioned as follows into a left-hand side and a right-hand side:

LeftHandSide ::= RightHandSide.

The main predicate of the compiler is translate_rule and relates a DCTG rule to a Prolog clause. The predicate translate_rule has two arguments typed as:

translate_rule(DCTG rule, Prolog clause)

It makes use of the following auxiliary predicates:

t_lp   translate left part of syntax

t_lp   translate right part of syntax

reverse   reverse a list

tidy   tidy up a conjunction of goals

There are a few special cases. If the right-hand side of the syntactic portion of the rule is empty, only the left-hand side of the DCTG needs to be processed:

```
translate_rule((LP::=[]<:>Sem),H) :- !,
  t_lp(LP,[],S,S,Sem,H).
```

If the right-hand side of the syntactic portion of the rule is empty and if no semantic rules are specified, processing is even simpler:

```
translate_rule((LP::=[]),H) :- !, t_lp(LP,[],S,S,[],H).
```

The most general form of DCTG rule requires translation of the right-hand side by *t_rp* which produces (among other things) a list of subtrees of the nonterminal symbol. This list is produced in reverese order (see below) and must be reversed. Then the left-hand side is translated by *t_lp*, and tidied up by tidy (see below):

```
translate_rule((LP::=RP<:>Sem),(H:-B)):- !,
  t_rp(RP,[],StL,S,SR,B1),
  reverse(StL,RStL),
  t_lp(LP,RStL,S,SR,Sem,H),
  tidy(B1,B).
```

The last clause of *translate_rule* takes care of the case when no semantic component is specified. An empty list is inserted as the semantic component and processing is recursively handled:

```
translate_rule((LP::=RP),(H:-B)):-
  translate_rule((LP::=RP<:>[]),(H:-B)).
```

The predicate t_lp

The predicate *t_lp* which translates the left-hand side of the syntactic portion of the rule has the following types of arguments:

```
t_lp( Left hand side, List of subtrees, Start list, End list,
      Semantics, Head of Prolog clause)
```

It makes use of the following auxiliary predicates:

append	append two lists
makelist	form a list from a conjunction
add_extra_args	add 3 arguments to a term

The first clause of *t_lp* deals with the case (briefly discussed earlier) in which a rule may be applied only within the context of a list of terminal symbols *List*. This is processed by appending *List* to *SR* which represents the second component of the difference list representing the input string. The conjunction of semantic rules is transformed into a list, and the predicate *add_extra_args* adds three arguments to the nonterminal symbol *LP:* the first is the representation used in creating the parse tree, and the last two represent the difference list for the string being analysed or synthesised. See below for the specification of these auxiliary predicates.

```
t_lp((LP,List),StL,S,SR,Sem,H) :-
  append(List,SR,List2),
  makelist(Sem,Semantics),
  add_extra_args([node(LP,StL,Semantics),S,List2],LP,H).
```

The other clause of *t_lp* which deals with the more usual case of DCTG rule transforms the conjunction of semantic rules and adds the extra arguments to the nonterminal symbol *LP*.

```
t_lp(LP,StL,S,SR,Sem,H) :-
  makelist(Sem,Semantics),
  add_extra_args([node(LP,StL,Semantics),S,SR],LP,H).
```

The Predicate t_rp

The predicate *t_rp* for translating the right hand side of the syntactic portion of a DCTG rule has the following types of arguments:

```
t_rp( Symbol, Subtrees in, Subtrees out, Start list, End list,
     Prolog equivalent of Symbol)
```

The list *Subtrees in,* initialized in *translate_rule* to the empty list may be extended by one or more symbols as *t_rp* proceeds. Some symbols which may appear in DCTG rules (such as the cut) do not introduce subtrees. The two components of the difference list *Start list* and *End list* are logical variables which at execution time represent the list of things which are analyzed as a *Symbol*.

t_rp: !

The symbol is ! which does not appear in the list of subtrees, does not use up any characters in the difference list, and is translated to the Prolog cut, i.e., it is left alone.

t_rp(!,St,St,S,S,!) :- !.

t_rp: [ ]

If the symbol is *[]*, this must be inserted as a subtree, and the Prolog equivalent of recognizing the empty string is to unify *S* and *S1*, the two components of the difference list:

t_rp([],St,[[]|St],S,S1,S=S1) :- !.

t_rp: One Character String

If the symbol is a single element string, that is, a character, because of the way C-Prolog deals with characters, the predicate *char* must convert the character (represented as an ASCII integer) to an atom. The atom corresponding to the character becomes one of the subtrees of the nonterminal symbol, and the Prolog program corresponding to the character is a call to the predicate *c*, specifying that the character is the difference of the two components of the list.

t_rp([X],St,[[NX]|St],S,SR,c(S,X,SR)) :-
char(X,NX).

t_rp: One Element List

If the symbol is a single element list, the element being a term, it is handled like the single character string except that the binary predicate *char* need not be called:

t_rp([X],St,[[X]|St],S,SR,c(S,X,SR)) :- !.

t_rp: String

If the symbol is a string, the first character is treated as above, and *t_rp* is recursively called to process the remaining string:

t_rp([X|R],St,[[NX|NR]|St],S,SR,(c(S,X,SR1),RB)) :-
char(X,NX),
t_rp(R,St,[NR|St],SR1,SR,RB).

t_rp: List

If the symbol is a list, the first term is treated as above, and *t_rp* is recursively called to process the remainder of the list:

t_rp([X|R],St,[[X|R]|St],S,SR,(c(S,X,SR1),RB)) :- !,

t_rp(R,St,[R|St],SR1,SR,RB).

t_rp: A Prolog Call

If the symbol is a call of a Prolog predicate, the enclosing braces are removed, the call is inserted in the clause under construction, and nothing is added to the list of subtrees of the terminal symbol.

t_rp({T},St,St,S,S,T) :- !.

t_rp: Conjunction of Symbols

If the symbol is a conjunction of DCTG symbols, there are two recursive calls of *t_rp*, one to handle the first element of the conjunction, the second for the other. The Prolog equivalent is the conjunction of the Prolog equivalents of the conjoined elements.

t_rp((T,R),St,StR,S,SR,(Tt,Rt)) :- !,
  t_rp(T,St,St1,S,SR1,Tt),
  t_rp(R,St1,StR,SR1,SR,Rt).

t_rp: Named Nonterminal Tree

If the symbol is a nonterminal T with tree name N, then a call to *add_extra_args* uses this variable N as the name of the subtree corresponding to nonterminal T. The two components of the difference list S and SR are also added to the term T.

t_rp(T^^N,St,[N|St],S,SR,Tt) :- add_extra_args([N,S,SR],T,Tt).

t_rp: Unnamed Nonterminal Tree

If the symbol is a nonterminal T but is not named, an arbitrary logical variable is used to name the subtree corresponding to the nonterminal T. The two components of the difference list S and SR are also added to the term T.

t_rp(T,St,[St1|St],S,SR,Tt) :- add_extra_args([St1,S,SR],T,Tt).

Auxiliary Predicates

add_extra_args

The predicate *add_extra_args* converts a term T to a list, appends L to it giving a new list $Tl1$ which is then converted to a term.

```
add_extra_args(L,T,T1) :-
 T=..Tl,
 append(Tl,L,Tl1),
 T1=..Tl1.
```

Append, Reverse

No comment should be necessary for the following:

```
append([],L,L) :- !.
append([X|R],L,[X|R1]) :- append(R,L,R1).
```

```
reverse(X,RX) :- rev1(X,[],RX).
```

```
rev1([],R,R) :- !.
rev1([X|Y],Z,R) :- rev1(Y,[X|Z],R).
```

Tidy

The predicate *tidy* adjusts the association of conjuctions.

```
tidy(((P1,P2),P3),Q) :-
 tidy((P1,(P2,P3)),Q).
```

```
tidy((P1,P2),(Q1,Q2)) :- !,
 tidy(P1,Q1),
 tidy(P2,Q2).
```

```
tidy(A,A) :- !.
```

Char

The predicate *char* converts a suitable integer to a readable constant symbol representing it.

```
char(X,NX) :-
 integer(X), X < 256, !, name(NX,[X]).
```

c

The predicate *c* specifies that *X* is the difference of the lists [*X/S*] and *S:*

```
c([X|S],X,S).
```

Using Term_Expansion

The following uses *term_expansion*, a system predicate of C-Prolog, to compile a DCTG rule T to a Prolog program clause E and to assert E.

```
:- asserta(( term_expansion(T,E) :- process_rule(T,E) , ! )).
```

Process_Rule

The predicate *process_rule* calls *translate_rule* and then asserts the DCTG rule T. Thus, one can list all DCTG rules by means of the system call:
```
:- listing(::=).
```

```
process_rule(T,E) :-
  translate_rule(T,E),
  !,
  assert(T).
```

Makelist

The predicate *makelist* takes converts the conjunction of semantic rules into a list.

```
makelist(Sem,[Sem]) :- var(Sem),!.

makelist((Sem1,Sem2),[Sem1|List]) :- !,
  makelist(Sem2,List).

makelist([],[]) :- !.

makelist(Sem,[Sem]).
```

The Operator ^^

Here are the definitions for the operator ^^. They define an interpreter for the clauses which define the semantic attributes.

```
node(_,_,Sem)^^Args :- Sem^^Args.

[(Args::-Traverse)|Rules]^^Args :-
  Traverse.

[Args|Rules]^^Args.

[_|Rules]^^Args :-
  Rules^^Args.
```

3.3. Typing DCTGs

Implementation Details

The adaptation of the representation of DCTG parse tree nodes and semantic rules to typing DCTGs is simple and straightforward. A node in a derivation tree which represents use of the production named *branch* in the grammar for trees is represented by the term:

    node(branch:tree,
        [[([], L, [,], R, [])]],
        [left(L), right(R)]).

In the DCTG node representation the first argument names only the non-terminal which is at the root of a derivation subtree; in a typing DCTG we also incorporate the name of the production used in forming that node. The second argument is a list of all subtrees of the node. Furthermore, the subtrees indicated above by L and R are node structures themselves which insure by unification that L and R are of type *tree* :

    L = node(N':tree, L', S')
    R = node(M':tree, R', T')

The semantic attributes of the *branch*:*tree* node are formed from the names of nonterminals in the right-hand side, or if the nonterminals in the right-hand side are labeled, from the labels; such attributes are unary function symbols which give access to the relevant subtrees of the node. Considered as Horn clauses for the semantic interpreter, they are unit clauses. Other semantic attributes to typing DCTG rules may be explicitly attached. If a nonterminal z is decorated with a logical variable X as in $z^{\wedge\wedge}X$, then X is instantiated to the derivation subtree for z, and the selector for z is compiled as $z(X)$. Semantic attribute rules (which go into the local data base for such a rule) may then traverse X to evaluate attributes, eg, $X^{\wedge\wedge}logic\,(A,B)$. Semantic rules may be specified as in DCTG rules following the <:> symbol. The generated attributes are appended to any such specified attributes to form the third argument of a node.

The method used to compile typing DCTG rules into Prolog clauses is a variant of the one used to compile DCTG rules into Prolog clauses. The differences are that:

1. Semantic typing attributes are formed and added to the list of other specified semantic attributes.

2. A predicate is formed from the name of each production and asserted.

3. Clauses for *type* and *attributes* are asserted for each typing DCTG rule.

Here is the way the grammar for natural numbers appears as Prolog clauses:

```
natural(node(zero:natural,[[0]],true),S0,S1) :-
  c(S0,0,S1).

natural(node(succ:natural,
      [[s,'('],node(Name:natural,Nodes,Semantics),[')']],
      [natural(node(Name:natural,Nodes,Semantics))]),
      S0,
      S4) :-
  c(S0,s,S1),
  c(S1,'(',S2),
  natural(node(Name:natural,Nodes,Semantics),S2,S3),
  c(S3,')',S4).
```

The first argument to *natural* is instantiated to the derivation tree, the second and third represent the difference list used in parsing. The following conjunction of goals shows how the generated Prolog parser *natural* of arity 3 may be used:

```
:- natural(N1,"s(s(0))",[]),
   natural(N2,"s(s(s(0)))",[]),
   sum(N1,N2,N3).
```

Further implementation details may be found in Abramson (1984b).

LITERATURE

Abramson, H.D. (1973) Theory and Application of a Bottom-up Syntax-Directed Translator. Academic Press.

Abramson, H. (1984) Definite Clause Translation Grammars. Proceedings of the IEEE Logic Programming Symposium, Atlantic City, pp. 233-240

Abramson, H. (1984) Definite Clause Translation Grammars and the Logical Specificaion of Data Types as Unambiguous Context Free Grammars. Proceedings of the International Conference on Fifth Generation Computer Systems, North-Holland, pp. 678-685.

Abramson, H. (1986) Sequential and Concurrent Deterministic Logic Grammars. In: E. Shapiro (ed) Third International Conference on Logic Programming, London, LNCS 225, Springer-Verlag.

Abramson, H. (1986) A Prological Definition of HASL, a Purely Functional Language with Unification-Based Conditional Binding Expressions. In: Logic Programming: Functions, Relations, and Equations, D. DeGroot and G. Lindstrom (eds), Prentice-Hall.

Abramson, H. (1988) Metarules and an Approach to Conjunction in Definite Clause Translation Grammars: Some Aspects of Grammatical Metaprogramming. In: Logic Programming, Proceedings of the Fifth International Conference and Symposium, R.A. Kowalski and K.A. Bowen (eds), pp. 233-248.

Abramson, H., Crocker, M., Ross, B., and Westcott, D. (1988) A Fifth Generation Translator Writing System. UBC Dept. of Computer Science Technical Report.

Aho, A.V., and Ullman, J.D. (1972) Theory of Parsing, Translation and Compiling, 2 vols. Prentice-Hall.

Aho, A.V., and Ullman, J.D. (1977) Principles of Compiler Design, Addison-Wesley.

Aho, A.V., Sethi, R., and Ullman, J.D. (1986) Compilers: Principles, Techniques, and Tools. Addison-Wesley.

Aida, H., Tanaka, H., and Moto-oka, T. (1983) A Prolog Extension for Handling Negative Knowledge. In: New Generation Computing, vol. 1, no. 1, pp. 87-91,

Bach, E. (1976) An Extension of Classical Transformational Grammar. University of Massachussetts, Amherst, MA.

Bates, M. (1978) The Theory and Practice of Augmented Transition Network Grammars. In: Natural Language Communication With Computers. L. Bolc (ed), Springer-Verlag, pp. 191-259.

Blair, H. (1982) The Undecidability of Two Completeness Notions for the "negation as failure" Rule in Logic Programming. First International Logic Programming Conference, Marseille, France, pp. 164-168.

Bratko, I. (1986) Prolog Programming for Artificial Intelligence, Addison-Wesley.

Brown, C., (1987) Generating Spanish Clitics using Static Discontinuity Grammar. PhD Thesis, School of Computing Sciences, Simon Fraser University.

Brown, C., Pattabhiraman T., Boyer M., Massam D., and Dahl V. (1986a) Tailoring Conceptual Graphs for Use in Natural Language Translation. Proceedings of the 1986 Thornwood Conference on Conceptual Graphs.

Brown, C., Dahl V., Massam D., Boyer M., and Pattabhiraman T. (1986b) Tailoring GB Theory for a Useful Subset of Error Messages. LCCR TR 86-4, Simon Fraser University, Burnaby, BC, Canada.

Brown, C., Pattabhiraman, T., and Massicotte P. (1988) Towards a Theory of Natural Language Generation: The Connection Between Syntax and Semantics. In: Natural Language Understanding and Logic Programming II, V. Dahl and P. Saint-Dizier (eds), North-Holland.

Chomsky, N. (1982) Lectures on Government and Binding, the Pisa Lectures 2nd edition. Foris Publications, Holland.

Clark, K.L. (1977) Negation as Failure. In: Logic and Data Bases, H. Gallaire and J. Minker (eds), pp. 293-324.

Clark, K.L. (1980) IC-Prolog-Language Features. Proceedings of the Logic Programming Workshop, Debrecen, Hungary.

Clark, K.L., Tärnlund, S.A.(1982) Logic Programming, Academic Press.

Clark, K.L., and Gregory, S. (1983) PARLOG: A Parallel Logic Programming Language. Research Report DOC 83/5, Dept. of Computing, Imperial College, London.

Clark, K.L., and Gregory, S. (1984) PARLOG: Parallel Programming in Logic. Research Report DOC 84/4, Dept. of Computing, Imperial College.

Clark, K.L., and MacCabe, F.G. (1984) Micro-PROLOG: Programming in Logic, Prentice-Hall.

Cleaveland, J.C., and Uzgalis, R.C. (1977) Grammars for Programming Languages. Elsevier.

Clocksin, W.F., and Mellish, C.S.(1981) Programming in Prolog, Springer-Verlag.

Codd, E.F. (1970) A Relational Model of Data for Large Shared Data Banks. Comm. ACM vol. 13, no. 6, pp. 377-397.

Coelho, H.M.F. (1979) A Program Conversing in Portuguese Providing a Library Service. Ph.D. Thesis, University of Edinburgh.

Cohen, J., and Hickey, T.J., (1987) Parsing and Compiling Using Prolog. ACM Transactions on Programming Languages and Systems. vol. 9, no. 2, pp. 125-163.

Colmerauer, A., et al. (1973) Un Systeme de Commumication Homme-Machine en Francais. Aix-Marseille University.

Colmerauer, A. (1978) Metamorphosis Grammars. In: L. Bolc (ed), Lecture Notes in Computer Science, Springer-Verlag, vol. 63, pp. 133-189.

Colmerauer, A. (1982) An Interesting Subset of Natural Language. In: Logic Programming, K. Clark and S.A. Tarnlund (eds), pp. 45-66.

Colmerauer, A., et al. (1983) Prolog, Theoretical Principles and Current Trends. North Oxford Academic/Afcet/Bordas, vol. 2, no. 4, pp. 255-292.

Crammond, J.A. (1986) An Execution Model for Committed-Choice Nondeterministic Languages. Proceedings of the 1986 Symposium on Logic Programming, IEEE, Salt Lake City, pp. 148-158.

Dahl, V., and Sambuc, R. (1976) Un Systeme de Banque de Donnees en Logique du Premier Ordre. en Vue de sa Consultation en Langue Naturelle, D.E.A. Report, Aix-Marseille University.

Dahl, V. (1977) Un Systeme Deductif d'Interrogation de Banques de Donnees en Espagnol. These de Doctorat de Specialite, Aix-Marseille University.

Dahl, V. (1979) Quantification in Three-Valued Logic for Natural Language Question-Answering Systems. Proceedings of the VI International Joint Conference on Artificial Intelligence , Tokyo.

Dahl, V. (1980a) Two Solutions for the Negation Problem. Proceedings of the Logic Programming Workshop, Debrecen, Hungary, pp. 61-72.

Dahl, V. (1980b) A Three-Valued Logic for Question-Answering Systems. Proceedings of the International Symposium on Multiple-Valued Logic, Illinois.

Dahl, V. (1981) Translating Spanish into Logic Through Logic. Americal Journal of Computational Linguistics, vol. 7, no. 3, pp. 149-164.

Dahl, V. (1982) On Database Systems Development Through Logic. ACM Transactions on Database Systems, vol. 7, no. 1, pp. 102-123.

Dahl, V. (1983) On Logic Programming as a Representation of Knowledge. In: IEEE Computer Special Issue on Knowledge Representation, pp. 106-111.

Dahl, V., and McCord, M. (1983) Treating Coordination in Logic Grammars. Americal Journal of Computational Linguistics, vol. 9, no. 2, pp. 69-91.

Dahl, V. (1984a) Logic Programming and Constructive Expert Systems. Proceedings of the First International Workshop on Expert Database Systems, South Carolina.

Dahl, V., and Abramson, H. (1984) On Gapping Grammars. Proceedings of the Second International Logic Programming Conference, Uppsala, pp. 77-88.

Dahl, V. (1984b) More on Gapping Grammars. Proceedings of the International Conference on Fifth Generation Computer Systems, Tokyo.

Dahl, V., and Saint-Dizier, P. (eds) (1985) Natural Language and Logic Programming, North-Holland.

Dahl, V. (1986) Gramaticas Discontinuas: Una Herramienta Computational con Aplicaciones en La Teoria de Reccion y Ligamiento. Revista Argentina de Linguistica, vol 2, no. 2, pp. 375-393.

Dahl, V., Brown, C., and Hamilton, S. (1986) Static Discontinuity Grammars and Logic Programming. LCCR TR 86-17, Simon Fraser University, Canada.

Dahl, V., and Saint-Dizier, P. (1986) Constrained Discontinuous Grammars-A Linguistically Motivated Tool for Processing Language. LCCR TR 86-8, Simon Fraser University, Burnaby, Canada.

Dahl, V. (1988a) Representing Linguistic Knowledge through Logic Programming. In: Proceedings V International Conference and Symposium on Logic Programming, Seattle.

Dahl, V. (1988b) Static Discontinuity Grammars for Government-Binding Theory. In: Proceedings Workshop "Informatique et Langue Naturelle", Nantes.

Dahl, V., and Massicote, P. (1988) Metaprogramming for Discontinuous Grammars. In: Proceedings of the Workshop on Meta-Programming in Logic Programming, Bristol.

Dahl, V., and Saint-Dizier, P. (eds) (1988) Natural Language Understanding and Logic Programming II, North-Holland.

Dahl, V., Levine, R., Miyoshi, H., Saint-Dizier, P., and Stabler, E.P.,Jr. (1988) Logic Grammar and Linguistic Theories, In: Natural Language and Logic Programming II, V. Dahl and P. Saint-Dizier (eds), North-Holland.

DeGroot, D., and Lindstrom, G. (eds) (1986) Logic Programming: Functions, Relations, and Equations. Prentice-Hall.

Deransart, P., and Maluszynski, J. (1984) Modelling Data Dependencies in Logic Programs by Attribute Schemata. INRIA-Rocquencourt, BP105 78153 Le Chesnay, France. Reseach Report No. 323, July.

Deransart, P., Jourdan, M., and Lorho, B. (1985) A Survey on Attribute Grammars: Part III Classified Bibliography. INRIA-Rocquencourt, BP105 78153 Le Chesnay, France. Reseach Report No. 510, June.

Deransart, P., Jourdan, M., and Lorho, B. (1986a) A Survey on Attribute Grammars: Part I Main Results on Attribute Grammars. INRIA-Rocquencourt, BP105 78153 Le Chesnay, France. Reseach Report No. 485, January.

Deransart, P., Jourdan, M., and Lorho, B. (1986b) A Survey on Attribute Grammars: Part II Review of Existing Systems. INRIA-Rocquencourt, BP105 78153 Le Chesnay, France. Reseach Report No. 510, March.

Earley, J. (1970) An Efficient Context-Free Parsing Algorithm. Communications of the ACM, vol. 13, pp. 94-102.

Filgueiras, M. (1986) Cooperating Rewrite Processes for Natural Language Analysis. The Journal of Logic Programming, vol. 3, no. 4.

Gallaire, H., Minker, J., and Nicolas, J.M. (1984) Logic and Databases: A Deductive Approach. ACM Computing Surveys, vol. 16, no. 2, pp. 153-186.

Gazdar, G. (1981) Unbounded Dependencies and Coordinate Structure. Linguistic Inquiry, vol. 12, no. 2, pp. 155-184.

Gazdar, G., Klein, E., Pullum, G.L., and Sag, I. (1985) Generalized Phrase Structure Grammar. Harvard University Press.

Giannesini, F., Kanoui, H., Pasero, R., and Van Caneghem, M. (1986) Prolog. Addison-Wesley.

Gregory, S.(1987) Parallel Logic Programming in PARLOG, Addison-Wesley.

Greibach, S. and Hopcroft, J. (1969) Scattered Context Grammars. Journal of Computer and System Sciences, No. 3, pp. 233-247.

Hausser, R. (1984) Quantification in an Extended Montague Grammar. Dissertation, University of Texas at Austin.

Hirschman, L. (1987) Conjunction in Meta-Restriction Grammar. The Journal of Logic Programming, vol.3, no. 4.

Hirschman, L., and Puder, K. (1982) Restriction Grammars in Prolog. Proceedings of the First International Logic Programming Conference, Marseille, pp. 85-90.

Hirschman, L., and Puder, K. (1984) Restriction Grammar: A Prolog Implementation. In: Logic Programming and its Applications, D.H.D. Warren and M. van Caneghem (eds.), Ablex.

Hogger, C.J.(1984) Introduction to Logic Programming, Academic Press.

Ingerman, P.Z. (1966) A Syntax-Oriented Translator. Academic Press.

Jaffer, J., Lassez, J.L., and Lloyd, J. (1983) Completeness of the Negation as Failure Rule. Proceedings of the International Joint Conference on Artificial Intelligence, pp. 500-506.

Johnson, M. (1987) The Use of Knowledge of Language. Brain and Cognitive Sciences, M.I.T.

Johnson, M. (1988) Move-α and the Unfold-Fold Transformation. Brain and Cognitive Sciences, M.I.T.

Joshi, A.K. (1985) Tree Adjoining Grammars: How much context-sensitivity is required to provide reasonable structural descriptions? In: Natural Language Parsing, Dowty et al. (eds), Reidel, pp. 206-250.

Kaplan, R.M. (1971) The MIND System: a Grammar Rule Language. The Rand Corporation, Santa Monica.

Kay, M. (1976) Experiments with a Powerful Parser. American Journal of Computational Linguistics, microfiche no 43.

Khabaza, T. (1984) Negation as Failure and Parallelism. Proceedings of the International Symposium on Logic Programming, pp. 70-75.

Knuth, D.E. (1968) Semantics of Context-Free Languages. Mathematical Systems Theory, vol.2, no. 2, pp. 127-145.

Knuth, D.E. (1971) Top Down Syntax Analysis. Acta Informatica, vol. 1, no. 2, pp 79-110.

Kowalski, R. (1974) Predicate Logic as a Programming Language. Proceedings of IFIP 74, North Holland, Amsterdam, pp. 569-574.

Kowalski, R. (1979) Logic for Problem Solving. North Holland.

Lloyd, J.W. (1987) Foundations of Logic Programming 2nd Edition. Springer-Verlag (1st Edition 1984).

Maluszynski, J. (1982) Towards a Programming Language Based on the Notion of Two-Level Grammar. Software Systems Research Center, Linköping Institute of Technology. Report LITH-MAT-R-82-16.

Maluszynski, J., and Nilsson, J.F. (1981) A Notion of Grammatical Unification Applicable to Logic Programming Languages. Dept. of Computer Science, Technical University of Denmark, Doc. ID 967.

Maluszynski, J., and Nilsson, J.F. (1982) A Version of Prolog Based on the Notion of Two-Level Grammar. Prolog Programming Environments Workshop, Linköping University, March.

Maluszynski, J., and Nilsson, J.F. (1982a) Grammatical Unification. Information Processing Letters vol. 15 no. 4. pp. 150-158.

Maluszynski, J., and Nilsson, J.F. (1982b) A Comparision of the logic programmming language Prolog with two-level grammars. Proceedings of the First International Logic Programming Conference, Faculte des Sciences de Luminy, Marseille. pp. 193- 199.

Maluszynski, J. (1982) Towards a Programming Language Based on the Notion of Two-Level Grammar. Software Systems Research Center, Linköping Institute of Technology. Report LITH-MAT-R-82-16.

Massicotte, P. and Dahl, V. (1988) Handling Concept-Type Hierachies through Logic Programming. Proc. Third Annual Workshop on Conceptual Graphs, in conjunction with AAAI-88.

Matsumoto, Y., Tanaka, H., Hirakawa, H., and Miyoshi, H. (1983) BUP: a Bottom-up Parser Imbedded in Prolog. New Generation Computing, vol. 1, no. 2, pp. 145-158.

McCord, M. (1981) Focalizers, the Scoping problem, and Semantic Interpretation Rules in Logic Grammars. Proceedings of the Workshop on Logic Programming for Intelligent Systems, Logicon Inc.

McCord, M. (1982) Using Slots and Modifiers in Logic Grammars for Natural Language. Artificial Intelligence, vol. 18, no. 3, 327-367.

McCord, M. (1984) Semantic Interpretation for the Epistle System. Proceedings of the Second International Logic Programming Conference, Uppsala, Sweden, pp. 65-76.

McCord, M. (1985) Modular Logic Grammars. Proceedings of the ACL Conference, July.

McCord, M., Dahl, V., and Abramson, H. (eds) (1986) Special Issue on Natural Language and Logic Programming. The Journal of Logic Programming, vol.3, No. 4.

McCord, M. (1986) Focalizers, the Scoping Problem, and Semantic Interpretation Rules in Logic Grammars. In: Logic Programming and Its Applications, D.H. Warren and M. van Caneghem (eds), Ablex.

Miyoshi, H. and K. Furukawa (1985) Object-Oriented Parser in the Programming Language ESP. In: Natural Language Understanding and Logic

Programming, V. Dahl and P. Saint-Dizier (eds), North-Holland.

Moss, C.D.S. (1980) A Formal Description of ASPLE Using Predicate Logic. DOC 80/18, Imperial College.

Moss, C.D.S. (1981) The Formal Description of Programming Languages using Predicate Logic. Ph.D. Thesis, Imperial College.

Moss, C.D.S. (1982) How to Define a Language Using Prolog. Conference Record of the 1982 ACM Symposium on Lisp and Functional Programming, Pittsburgh, Pennsylvania. pp. 67-73

Naish, L.(1982) An Introduction to MU-PROLOG, Technical Report 82/2, Dept. of Computer Science, University of Melbourne.

Naish, L. (1983) An Introduction to Mu-Prolog, Technical Report. University of Melbourne, pp. 1-16.

Naish, L. (1985) Negation and Control in Prolog. Lecture Notes in Computer Science, vol. 238. Springer-Verlag.

Nilsson, U. (1986) AID: An Alternative Implementation of Definite Clause Grammars, New Generation Computing, vol. 4, pp. 383-395.

Pereira, L.M., Pereira, F.C.N., and Warren, D.H.D. (1978) User's Guide to Decsystem-10 Prolog, Technical Report. University of Edinburgh, pp. 1-60.

Pereira, F.C.N., and Warren, D.H.D. (1980) Definite Clause Grammars for Language Analysis − A Survey of the Formalism and a Comparison with Transition Networks. Artificial Intelligence, vol. 13, pp. 231-278.

Pereira, F.C.N. (1981) Extraposition Grammars. American Journal of Computational Linguistics, vol. 9, no. 4, pp. 243-255.

Pereira, F. (1983) Logic for Natural Language Analysis. Technical Note 275, SRI International.

Pereira, F.C.N. (ed)(1984) C-Prolog User's Manual, Version 1.5.

Pereira, F.C.N. and Shieber, S.M. (1987) Prolog and Natural-Language Analysis. CSLI Lecture Notes #10.

Pique, J.F. (1982) On a Semantic Representation of Natural Language Sentences. Proceedings of the First International Logic Programming Conference, Marseille, pp. 215-223.

Popowich, F. (1985) Unrestricted Gapping Grammars, Proc. IJCAI-85

Porto, A., and Filgueiras, M. (1984) Natural Language Semantics: A Logic Programming Approach. International Symposium on Logic Programming, Atlantic City, pp. 228-232.

Pratt, V. (1975) LINGOL, A Progress Report. Advance Papers, 4th International Joint Conference on Artificial Intelligence, Tbilisi, Georgia, USSR. pp. 422-428.

Pullum G.K. (1982) Free Word Order and Phrase Structure Rules. In: Proc 12th Annual Meeting North Eastern Linguistic Society, J.Pustejovsky and P.Sells, (eds).

Reiter, R. (1977) On Closed-World Databases. In: Logic and Databases, Gallaire and Minker (eds.), pp. 55-76.

Robinson, J.A. (1965) A Machine-Oriented Logic Bases on the Resolution Principle. Journal of the ACM 12, pp. 23-44.

Robinson, J.A. (1979) Logic: Form and Function. University Press, Edinburgh.

Ross, J.R. (1974) Excerpts from Constraints on Variables in Syntax. In: On Noam Chomsky, Critical Essays, Anchor Press.

Roussell, P. (1975) Prolog: Manuel de Reference et d'Utilisation. Aix-Marseille University.

Sabatier, P. (1985) Quantifier Hierarchy in a Semantic Representation of Natural Language Sentences. In: Natural Language Understanding and Logic Programming, V. Dahl and P. Saint-Dizier (eds), North-Holland.

Sag, I.A., (1982) A Semantic Theory of NP-movement Dependencies, In: The Nature of Syntactic Representation, P. Jacobson and G.K. Pullum (eds), Reidel.

Sager, N. (1967) Syntactic Analysis of Natural Language. In Advances in Computers. Academic Press.

Sager, N. (1981) Natural Language Information Processing. Addison-Wesley.

Sager, N., and Grishman, R. (1975) The Restriction Language for Computer Grammars of Natural Language. Communications of the ACM, vol. 18. pp. 390-400.

Saint-Dizier, P. (1986) An Approach to Natural-Language Semantics in Logic Programming. The Journal of Logic Programming, vol. 3, no. 4.

Saint-Dizier, P. (1985) Handling Quantifier Scoping Ambiguities in a Semantic Representation of Natural Language Sentences. In: Natural Language Understanding and Logic Programming, V.Dahl and P.Saint-Dizier (eds.), North-Holland.

Saint-Dizier, P. (1987) DISLOG: Programming in Logic with Discontinuities. LCCR TR 87-13, Simon Fraser University, Canada.

Saint-Dizier, P. (1988) Contextual Discontinuous Grammars. In: Natural Language Understanding and Logic Programming II, V. Dahl and P. Saint-Dizier (eds), North-Holland.

Sedogbo, C. (1985) A Meta-Grammar for Handling Coordination in Logic Grammars. In: Natural Language Understanding and Logic Programming, V. Dahl and P. Saint-Dizier (eds.), North-Holland.

Shapiro, E. (1983) A Subset of Concurrent Prolog and Its Interpreter. ICOT Technical Report, TR-003, Tokyo.

Sowa, J. (1984) Conceptual Structures. Addison-Wesley.

Stabler, E. (1983) Deterministic and Bottom-up Parsing in Prolog. Proceedings of AAAI-83, pp. 383-386.

Stabler, E.P., Jr. (1987) Restricting Logic Grammars with Government-Binding Theory. To appear in: Journal of Computational Linguistics.

Stabler, E.P., Jr. (1988) Parsing with Explicit Representations of Syntactic Constraints. In: Natural Language Understanding and Logic Programming II, V. Dahl and P. Saint-Dizier (eds) North-Holland.

Sterling, L., and Shapiro, E. (1986) The Art of Prolog, MIT Press.

Thom, J.A., and Zobel, J. (1986) NU-Prolog Reference Manual, Version 1.1, Technical Report 86/10, Dept. of Computer Science, University of Melbourne.

Ueda, K. (1985) Guarded Horn Clauses. ICOT Tech. Report. TR-103, Institute for New Generation Technology.

Uehara, K., Ochitani, R., Kakusho, O., and Toyoda, J. (1984) A Bottom-up Parser Based on Predicate Logic: A Survey of the Formalism and its Implementation Technique. Proceedings of the 1984 International Symposium on Logic Programming, Atlantic City, pp. 220-227.

van Emden, M. (1977) Programming with Resolution Logic. In: Machine Intelligence 8, E. Elcock and D. Michie, (eds) Ellis Horwood.

Walker, A., McCord, M., Sowa, J.F., and Wilson, W.G. (1987) Knowledge Systems and Prolog. Addison-Wesley.

Wallace, M. (1984) Communicating with Databases in Natural Language. Ellis Horwood Services (A-1), European Computer-Industry Research Center GMBH.

Warren, D.H.D. (1977) Logic Programming and Compiler Writing. DAI Research Report 44, University of Edinburgh.

Warren, D.H.D. (1980) Logic Programming and Compiler Writing. Software Practice and Experience. vol. 10, pp. 97-125.

Warren, D.H.D. (1981) Higher-Order Extensions to Prolog – Are They Needed? Tenth International Machine Intelligence Workshop, Case Western Reserve University, Cleveland, Ohio.

Warren, D.H.D. (1981b) Efficient Processing of Interactive Relational Database Queries Expressed in Logic. Department of AI, University of Edinburgh.

Warren, D.H.D. (1982) An Efficient Easily Adaptable System for Interpreting Natural Language Queries. American Journal of Computational Linguistics, vol. 8, no. 3-4, pp. 110-119.

Warren, D.S. (1983) Using λ-calculus to Represent Meaning in Logic Grammars. Proceedings of the 21st Annual Meeting of the Association for Computational Linguistics, MIT, Cambridge, MA, pp. 51-56.

Winograd, T. (1972) Understanding Natural Language. Academic Press.

Woods, W.A. (1973) An Experimental Parsing System for Transition Network Grammars. In: Natural Language Processing, R. Rustin (ed), Algorithmic Press.

INDEX